SURFACE PHYSICS
OF MATERIALS

VOLUME II

MATERIALS SCIENCE AND TECHNOLOGY

EDITORS

ALLEN M. ALPER

GTE Sylvania Inc.
Precision Materials Group
Chemical & Metallurgical
Division
Towanda, Pennsylvania

JOHN L. MARGRAVE

Department of Chemistry
Rice University
Houston, Texas

A. S. NOWICK

Henry Krumb School
of Mines
Columbia University
New York, New York

SURFACE PHYSICS OF MATERIALS

Edited by J. M. BLAKELY

*Department of Materials Science and
Engineering and Materials Science Center
Cornell University
Ithaca, New York*

VOLUME II

ACADEMIC PRESS   New York   San Francisco   London   1975

A Subsidiary of Harcourt Brace Jovanovich, Publishers

PHYSICS

ACADEMIC PRESS, INC.
111 Fifth Avenue, New York, New York 10003

United Kingdom Edition published by
ACADEMIC PRESS, INC. (LONDON) LTD.
24/28 Oval Road, London NW1

Library of Congress Cataloging in Publication Data
Main entry under title:

Surface physics of materials.

(Materials science and technology series)
Includes bibliographies and indexes.
1.   Surfaces (Technology)   I.   Blakely, John McDonald.
TA418.7.S92   531   74-27774
ISBN 0–12–103802–5 (v. 2)

PRINTED IN THE UNITED STATES OF AMERICA

Contents

9. Surface Vibrations

M. G. Lagally

10. Interaction between Surfaces: Adhesion and Friction

D. Tabor

List of Contributors

Numbers in parentheses indicate the pages on which the authors' contributions begin.

H. P. BONZEL (279),* Exxon Research and Engineering Company, Corporate Research Laboratory, Linden, New Jersey

J. W. GADZUK (339), National Bureau of Standards, Washington, D. C.

M. G. LAGALLY (419), College of Engineering, University of Wisconsin, Madison, Wisconsin

ROBERT L. PARK (377),† Sandia Laboratories, Albuquerque, New Mexico

D. TABOR (475), Department of Physics, Cavendish Laboratory, University of Cambridge, Cambridge, England

* Present address: Institut für Grenzflächenforschung und Vakuumphysik der KFA, 517 Jülich 1, West Germany.
† Present address: Department of Physics, University of Maryland, College Park, Maryland.

Preface

The material in these two volumes provides an up-to-date account of our understanding of the physical properties of solid surfaces. Research in this area has already attained a level of considerable sophistication, and surface science promises to continue to be an exciting and worthwhile field of endeavor in the foreseeable future. The various chapters critically examine the status of work on a number of aspects of solid surfaces and attempt to predict the most profitable avenues for future research. The spectacular increase in interest in surface physics, occurring in the last decade, has been sparked by the realization of the importance of surfaces and interfaces in solid state devices and chemical reactions. It is therefore an area of applied research; yet, as witnessed by numerous examples in this book, it is one that requires the techniques of both the materials engineer and the mathematical physicist.

The two volumes contain a set of papers carefully selected to give broad coverage of the field of surface physics. The individual chapters deal with topics of current research interest and have been chosen to emphasize surface properties rather than the applicability of experimental techniques. It is hoped that these volumes will be especially useful to research workers, teachers, and graduate students in surface physics as well as serving as reference texts for the materials scientist specializing in other branches of the subject.

The authors of the various chapters are all individuals who have made substantial contributions to the development of the particular areas about which they have written. In most cases emphasis has been placed on fundamentals and on those aspects that are least likely to require revision as the subject develops. I have found that editing this work has been an educational experience and can only hope that the reader will derive comparable benefits.

I am especially grateful to Aggie Sirrine, Marsha Leonard, and Karen Pratt for their assistance in preparing this volume for publication.

Contents of Volume I

SURFACE PHYSICS
OF MATERIALS

VOLUME II

6

Transport of Matter at Surfaces

H. P. BONZEL[†]

EXXON RESEARCH AND ENGINEERING COMPANY
CORPORATE RESEARCH LABORATORY
LINDEN, NEW JERSEY

[†] Present address: Institut für Grenzflächenforschung und Vakuumphysik der KFA, 517 Jülich 1, West Germany.

I. INTRODUCTION

Diffusion on crystalline surfaces is a process that has received much attention from a scientific point of view and that at the same time is also of considerable significance in a number of technological areas. In this chapter we are not going to discuss these related technological areas in any detail, but we would like to point out some of them to the uninitiated reader, together with a few key references. For example, the sintering of metal and metal oxide powders is to a large extent controlled by surface diffusion (Kuczynski, 1961; Rhead, 1972); the rate of migration and coalescence of internal voids in metals, as they are generated in nuclear reactors, is governed by surface diffusion (Willertz and Shewmon, 1970); the stability of small catalyst particles that are supported on high surface area oxides may be a function of surface diffusion of adatoms across the support (Geguzin *et al.*, 1969; Wynblatt and Gjostein, 1974); the process of crystal growth from the vapor phase has been recognized to depend on surface diffusion of adatoms (Hirth and Pound, 1963; Gilmer and Bennema, 1972); the technology of thin film semiconductor devices is in several ways related to the process of surface diffusion, e. g., through nucleation and growth of epitaxial films (e. g., Hayek and Schwabe, 1972; Bauer and Poppa, 1972) and electromigration in thin films (Rosenberg, 1972). Thus there is ample reason for carrying out research on surface diffusion if one should be concerned with the technological relevance of such research.

We define surface diffusion as the motion of individual entities, such as atoms, ions, molecules, and small clusters of atoms, across crystalline solid surfaces. This motion is assumed to be thermally activated and to take place in the adsorbed state, i.e., entirely "on top" of the free surface of the crystal. In particular, surface vacancies will be treated as an adsorbed diffusing defect.

This definition of surface diffusion implies that the so-called "thickness of the surface layer" in which the process of surface diffusion takes place is equal to one interplanar distance (of the order of 2×10^{-8}cm). Therefore we will not consider averaging a surface diffusion coefficient over several layers, as Rusanov (1971) proposed for certain other surface properties. In most crystalline solids lattice relaxation is not expected to be so large that a significant variation in the density of the solid would result as a function of the distance from the surface. For example, calculations of lattice relaxation by Burton and Jura (1967a) and Wynblatt and Gjostein (1968) for copper surfaces show that at most a 20% decrease in density between the two outermost layers is found for a Cu(110) surface. This means that the transition from solid to vacuum is very abrupt indeed.

The process of surface diffusion is strongly linked to the crystallographic nature of the surface and to the existence of a two-dimensional surface potential, very similar to the one first discussed by Lennard–Jones (1932). At first we will review the atomic models of crystalline surfaces. In this context we will also introduce surface defects generally assumed to be responsible for surface diffusion and discuss their energetics. The theory of surface diffusion coefficients is presented in the framework of statistical thermodynamics and is based on the pioneering ideas of Lennard–Jones (1932, 1937). Calculations of surface diffusion coefficients will be demonstrated where the required input are crystallographic data and energetic and entropic parameters.

In order to measure surface diffusion coefficients D_s, it is necessary to derive phenomenological relationships relating D_s to microscopic or macroscopic measurable parameters, such as mean diffusion distance, geometric parameters (particle size, grain boundary groove, amplitude of sine wave, etc.), concentration gradients (tracer diffusion), and time. These relationships will be discussed in a separate section.

There is a large variety of experimental techniques for measuring surface diffusion on crystalline solids. In this review we will only touch slightly on experimental questions and refer to previous works (Gjostein, 1970a; Bonzel, 1973) as well as original papers referenced elsewhere (see Table I).

TABLE I

RECENT REVIEWS OF SURFACE DIFFUSION

Title	Authors
Surface structure and diffusion	Gomer (1959)
Surface self-diffusion	Gjostein (1963)
Surface diffusion	Blakely (1963)
Surface self-diffusion in fcc and bcc metals: A comparison of theory and experiment	Gjostein (1967)
Zur Oberflächendiffusion und oberflächennahen Diffusion auf Kristallen	Meyer (1968)
Diffusion along a real crystal surface	Geguzin (1969)
In situ measurements of surface self-diffusion of metals	Bonzel and Gjostein (1969)
Surface diffusion of oxides	Robertson (1969)
Surface self-diffusion on metals	Hirano and Tanaka (1970)
Surface, grain boundary, and dislocation pipe diffusion	Gjostein (1970a)
Regularities of surface diffusion	Gal and Borisov (1971)
Surface self-diffusion of metals	Neumann and Neumann (1972)
Mobility of atoms and molecules over solid surfaces	Geuss (1972)
Surface diffusion of metals	Bonzel (1973)
Short circuit diffusion	Gjostein (1973)

From the very extensive literature of surface diffusion experiments, some examples will be presented in order to illustrate the enormous breadth of the phenomenon. Only a minor attempt at unifying the various results into a single picture will be undertaken, because our understanding of the microscopic processes involved, as well as the control of experimental conditions, is still quite limited. It will become apparent that the rate of surface diffusion is strongly dependent on the chemical composition of the surface and that therefore a wide spectrum of diffusivities is generally observed, even for a single material at a given temperature.

Finally it should be recognized that numerous reviews and summarizing articles on surface diffusion are available in the literature, which should be consulted for further information on specific questions. In order to facilitate this, a list of recent reviews has been compiled in Table I.

II. DEFECT MODEL OF CRYSTALLINE SURFACES

Considerable evidence has been advanced in recent years from low-energy electron diffraction (LEED) studies of single crystal surfaces (Strozier et al., in Chapter 1, Volume I of this work; Sickafus and Bonzel, 1971) that the structure of a surface results from a simple termination of the bulk crystallographic structure. This is considered to be the case for most metals, metal oxides, and some metal sulfides (Benard, 1969), whereas exceptions are found for certain orientations of Si, Au, Pt, and Ir (Somorjai, 1972). Other LEED experiments have shown that vicinal surfaces of UO_2, Cu, Si, and Pt give rise to diffraction features consistent with the presence of ordered arrays of monatomic steps (Ellis and Schwoebel, 1968; Perdereau and Rhead, 1969, 1971; Henzler, 1970a, b; Lang et al., 1972). Steps on crystalline surfaces have also been made visible by electron microscopy after heavy atoms were allowed to adsorb. This technique of decorating steps was originated by Bassett (1958) and has been used widely to characterize surfaces during or after evaporation (Bethge, 1969).

These experimental findings have thus far supported the terrace–ledge–kink (TLK) model of a crystalline surface that was conceived independently by Stranski (1928) and Kossel (1927). Figure 1 illustrates schematically the important features of this model, which was originally proposed for the simple cubic structure of a NaCl crystal surface. It is now widely accepted in the theory of surfaces and crystal growth (Hirth and Pound, 1963). The essential features of a TLK surface are atomically flat terraces separated by monatomic ledges (or steps) which themselves may exhibit kinks. Two other important defects exist on the surface: adatoms (adsorbed atoms) and terrace vacancies. These defects may also be adsorbed at a ledge and thus form a ledge-adsorbed atom and a ledge vacancy, respectively.

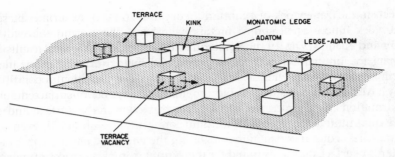

Fig. 1. Schematic diagram of a terrace–ledge–kink model of a single crystalline surface (Gjostein, 1967).

The number of ledges and kinks on a single crystal surface is related to the geometry of the surface as well as thermal fluctuations (Leamy *et al.*, Chapter 3, Volume I). For example, if a (100) surface of a simple cubic crystal is misoriented in [100] and [010] directions, then the average spacing between the ledges s_1 and the average kink density a_2 is given by the two misfit angles θ_1 and θ_2:

$$s_1 = \frac{h}{\tan\theta_1}, \qquad s_2 = \frac{h}{\tan\theta_2}, \tag{1}$$

where h is the height of a monatomic step and also the width of a kink.

In thermodynamic equilibrium the concentrations of surface defects, except those that exist for geometric reasons, are fixed and a function of temperature only (pressure dependence is generally neglected). This is particularly so for adatoms, terrace vacancies, ledge-adsorbed atoms, and (thermal) kinks (Burton *et al.*, 1951). In a qualitative sense it can be seen that the most simple defects of a TLK surface are adatoms and terrace vacancies. Their binding energies to the surface are lower than for all other defects as can be shown by a simple bond-counting method, and their mobility on the surface will likely be larger than that for other defects (at least over a large temperature range). Under such conditions their contribution to the process of surface diffusion will be overwhelming. Thus it is generally believed that surface diffusion takes place by the motion of adatoms and/or terrace vacancies, with some exceptions that we will discuss later.

III. ENERGETICS OF SURFACE DEFECTS

In order to develop a quantitative approach to the theory of surface diffusion, a detailed knowledge of the behavior of surface defects is needed. It would be desirable to calculate the fundamental properties of surface defects in the framework of quantum theory, but so far only surface energies and work functions of simple metals have been obtained in this way by Lang and Kohn (1970, 1971). Some quantum mechanical formulations describing the interaction between adatoms and metal surfaces are available (Grimley, 1969; Schrieffer, 1972; Gadzuk, Chapter 7 of this volume), but they have not yet been successful in providing data for energies and entropies of formation and migration of surface defects. However, the latter are needed for a quantitative theory of surface diffusion.

A less rigorous approach for calculating the energetic and entropic parameters of surface defects is based on the assumption of pairwise interactions of atoms in a solid. This approach, which was already exercised by Kossel (1927) and Stranski (1928), has provided much insight into the atomistics of surface diffusion. Many researchers have calculated the energetics of surface defects on the basis of a 6–12 Lennard–Jones or a Morse potential (Volmer, 1939; Neumann, 1938; Stranski and Suhrmann, 1947; Müller, 1949; Drechsler, 1954; Neumann et al., 1966; Goodman, 1967; Wynblatt and Gjostein, 1968, 1970; Ehrlich and Kirk, 1968; Burton and Jura, 1967b).

Wynblatt and Gjostein (1968, 1970) reported an improved method for calculating defect energies by using a Girifalco–Weizer (1959) potential and considering the lattice relaxation around defects. The potential is a modified Morse function

$$\varphi(r_{ij}) = A\{\exp[-2\alpha(r_{ij}-r_0)] - 2\exp[-\alpha(r_{ij}-r_0)]\}, \qquad (2)$$

where r_{ij} is the separation between two atoms i and j and α, r_0 and A are constants. According to Girifalco and Weizer (1959), these constants are determined by a set of boundary conditions involving the energy of sublimation, compressibility, and lattice constant of the solid.

Wynblatt and Gjostein (1968, 1970) carried out calculations of defect energies on the basis of this potential for Cu and W crystals containing about 3400 atoms. For the calculation of the formation energy of an adatom the procedure is as follows: An atom is removed from a kink site to infinity and then placed onto a terrace to form an adatom. The lattice in the immediate vicinity of the adatom is allowed to relax, yielding an energy ΔE_{AR}. Thus the formation energy of an adatom is

$$\Delta H_f^a = \Delta E_K - \Delta E_A - \Delta E_{AR}, \qquad (3)$$

with ΔE_K is the energy of a kink site and ΔE_A is the energy of an unrelaxed adatom.

Quite similarly, the energy of formation of a surface vacancy is obtained by removing an atom from a terrace site to infinity (energy contribution ΔE_T) and replacing it at a kink site (contribution ΔE_K). Then the lattice around the created terrace vacancy is permitted to relax (contribution ΔE_{VR}).

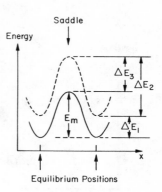

Fig. 2. Potential energy diagram illustrating the origin of the various energy terms that have to be computed in relationship to the migration energy $E_m \equiv H_m$ of a defect. The solid line represents the actual potential barrier for a jump, while the dashed line represents the potential barrier for the hypothetical jump (Wynblatt and Gjostein, 1968)

Finally one obtains

$$\Delta H_f^v = \Delta E_T - \Delta E_K - \Delta E_{VR}. \tag{4}$$

The calculation of migration energies is more difficult because of the involved saddle point positions. Figure 2 shows schematically the changes in energy accompanying a diffusion jump of an adatom, for instance, from one equilibrium position to another. The solid line represents the actual energy changes during the diffusion jump, where ΔH_m is the height of the diffusion barrier or the migration energy. The dashed line, on the other hand, represents a hypothetical energy path for conditions where all atoms except the jumping adatom are held at fixed positions. Thus for the dashed line the lattice is not necessarily relaxed. The energy ΔE_2 is calculated as the barrier height of the hypothetical potential. The energy ΔE_1 is equal to the relaxation energy around a normal lattice site (with respect to the nonequilibrium site from which the hypothetical jump occurred). ΔE_3 represents the relaxation energy of the saddle point, which is obtained by relaxing the atom at the saddle point in order to minimize the energy of that configuration. From Fig. 2 it is then easy to see that

$$\Delta H_m = \Delta E_1 + \Delta E_2 - \Delta E_3. \tag{5}$$

It should be noticed that the position of the minima of the dashed curve in Fig. 2 is rather arbitrary depending on where the adatom is placed on the surface for the calculation of ΔE_1 and ΔE_2. The starting configuration chosen by Wynblatt and Gjostein was an actual lattice site, while the underlying lattice was fully relaxed; i.e., the first few layers near the surface are displaced from actual lattice plane positions. The results of calculations for various surfaces of copper are summarized in Table II.

Wynblatt (1969) also calculated migration and formation entropies of surface defects for copper. These calculations were carried out for the same equilibrium crystal as before and by using the same potential. The crystal is

TABLE II

CALCULATED DEFECT ENERGIES FOR COPPER[a]

Defect	Surface	ΔH_f kcal/mole	ΔH_m kcal/mole	Q_s kcal/mole
Vacancy	(100)	11.3	3.9	15.2
Vacancy	$(110)\langle110\rangle$	11.7	7.0	18.7
Vacancy	(111)	19.3	15.1	34.4
Adatom	(100)	23.5	5.7	29.2
Adatom	$(110)\langle110\rangle$	13.8	1.36	15.2
Adatom	(111)	22.8	≈ 0.53–0.74	≈ 23.4

[a] From Wynblatt and Gjostein, 1968.

treated as an "Einstein solid," where each atom is characterized by three normal mode frequencies. The formation entropy of an adatom, for example, is then defined as

$$\frac{\Delta S_f}{k} = \ln\left(\frac{\prod_{j=1}^{3N} v_j^0}{\prod_{j=1}^{3N} v_j}\right), \tag{6}$$

where v_j^0 are the normal mode frequencies of the perfect reference crystal and the v_j frequencies of the crystal with one adatom adsorbed in an equilibrium position. Similarly, the migration entropy as derived by Vineyard (1957) is

$$v_e \exp\left(\frac{\Delta S_m}{k}\right) = \frac{\prod_{j=1}^{3N} v_j}{\prod_{j=1}^{3N-1} v_j^*}, \tag{7}$$

where v_j^* are now the frequencies for the crystal with the defect in a (relaxed) saddle point position. Or, if the equilibrium configuration of the defect is constrained in one direction such that it vibrates only in directions orthogonal to the jump direction, then

$$\frac{\Delta S_m}{k} = \ln\left(\frac{\prod_{j=1}^{3N-1} v_j'}{\prod_{j=1}^{3N-1} v_j^*}\right), \tag{8}$$

where the v_j' are the frequencies for the constrained state.

The normal mode frequencies v_j^0, v_j', and v_j^* were calculated by expanding the interatomic potential in a Taylor series about the equilibrium position of an arbitrary atom i. For small displacements along the Cartesian axes one obtains a harmonic approximation of the potential. The equations of motion for the oscillating defect vibrating in this potential can then be solved and the normal mode frequencies extracted.

The results of Wynblatt's computations, which are summarized in Table

III, indicate a D_0 factor in the range 2.5×10^{-3} to 3×10^{-2} cm^2/sec where

$$D_0 = al^2 v_e \exp\left(\frac{\Delta S_f + \Delta S_m}{k}\right). \tag{9}$$

These values are quite reasonable compared to extrapolated low temperature surface diffusion data of metals. However, one should be aware of some of the problems that arise in calculating entropies in the described manner. For a discussion of these problems one should consult Wynblatt's (1969) original paper.

TABLE III

CALCULATED DEFECT ENTROPIES FOR COPPER[a]

Defect	Surface	$\Delta S_f/k$	$\Delta S_m/k$	D_0(cm^2/sec)
Vacancy	(100)	2.82	0.095	3.38×10^{-2}
Vacancy	(111)	1.04	0.056	1.51×10^{-2}
Vacancy	(110)	0.58	0.19	2.45×10^{-3}
Adatom	(100)	1.46	0.28	9.28×10^{-3}
Adatom	(111)	2.45	-0.24	2.49×10^{-3}
Adatom	(110)	0.90	1.15	6.17×10^{-3}

[a] From Wynblatt, 1969.

Besides some specific problems with the calculation of surface energetic and entropic parameters, there is also a general problem: that of the interatomic potential. The Morse potential was derived through a matching process with *bulk* properties, but it was used to calculate *surface* properties. Furthermore, a Morse potential describes essentially central forces, whereas metal bonding is characterized to a large extent by noncentral forces. Hence it is not too surprising that some calculations such as the binding energy calculations for tungsten surfaces (Wynblatt and Gjostein, 1970) show poor agreement with experimental data despite included lattice relaxation. It is expected that quantum mechanical calculations will eventually lead to a clarification of these issues.

IV. STATISTICAL THERMODYNAMICS OF SURFACE DIFFUSION COEFFICIENTS

The process of surface diffusion is intimately connected with the crystallinity of solid surfaces and thus with the existence of surface sites. As we have already seen, these various surface sites are energetically different in such a way that the potential energy across a TLK surface is a complicated three-dimensional function. Accordingly, the diffusion of surface defects across this surface cannot proceed at a uniform rate. In order to calculate meaningful diffusion coefficients for such a complicated situation, we first look at the rate of surface diffusion on an atomically flat terrace (intrinsic diffusion coefficient). This diffusion is analogous to the diffusion of vacancies (or interstitials) in the bulk. In a second step we equilibrate the diffusing defects with defects at the boundary of a terrace, such as kinks, ledges, ledge-adsorbed atoms, ledge vacancies, etc. Through this equilibration a rate of exchange of defects between ledges and terraces is defined (dynamic equilibrium), and thus the continuity of surface diffusion across energetically less favorable sites is guaranteed. This diffusivity is called mass transfer diffusivity and is analogous to self-diffusion of atoms (vacancy mechanism) in the bulk. Because the time constant for the rate of exchange with tightly bound defects is larger than the time constant for intrinsic surface diffusion, mass transfer diffusion is slower than intrinsic diffusion of defects.

Even though surface and bulk diffusion are analogous in many respects, there are also some definite differences. It seems worthwhile to point out some of these differences before we proceed with the derivation of expressions for surface diffusion coefficients. We take the example of an adatom on a (100) terrace of a fcc metal, shown schematically in Fig. 3a. In order for this adatom to diffuse from one equilibrium site to a neighboring site, it must accumulate enough thermal energy through phonon interactions that it can surmount the potential energy barrier separating the two sites. A minimum energy path is along the $\langle 110 \rangle$ direction and is labelled (1) in Fig 3a. This path is across a saddle point whose energy we call the energy of

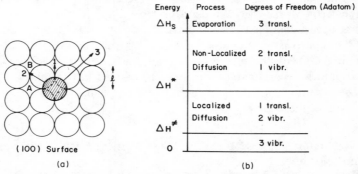

Fig. 3. (a) Assumed equilibrium position of adatom on a (100) oriented surface of a fcc metal. Adatom can diffuse locally to position 1 or A or B, depending on its momentum, and nonlocally to a hypothetical position 3. (b) Energy diagram for adatom illustrating the regimes of localized and nonlocalized diffusion as well as evaporation. Each state has a characteristic distribution of degrees of freedom. ΔH_s is here the binding energy of adatom to a terrace.

migration $\Delta H^{\ddagger}$ (Bonzel, 1970). The diffusing jump will probably be of the order of the interatomic distance of the lattice. On the other hand, if the same adatom accumulates much more energy than $\Delta H^{\ddagger}$, the minimum energy required for surface diffusion (tunneling excluded), then it is conceivable that it will be able to jump across the top of one of its nearest neighbors, indicated by path (3) in Fig. 3a. In that case it is less likely to interact with the surface, and consequently its flight path might be much longer than an interatomic distance. The minimum energy for this jump is ΔH^{*}, which is larger then $\Delta H^{\ddagger}$ but smaller than ΔH_{s}, the binding energy of an adatom on a terrace.

The energy scale is indicated in Fig. 3b. If the energy of an adatom is between $\Delta H^{\ddagger}$ and ΔH^{*}, we speak of localized diffusion; if it is in the range $\Delta H^{*} \leqslant \Delta H < \Delta H_{s}$, we call the diffusion "nonlocalized." This surface diffusion mode is characterized by a lower probability of deactivation and long jump distances (Bonzel, 1970). With the introduction of nonlocalized surface diffusion we can see an important difference between surface and bulk diffusion, namely that at a surface the diffusing defect can escape partially into the free space above the solid as long as its energy is less than its binding energy but larger than ΔH^{*}, whereas in the bulk there is no room for such "extra freedom."

According to the definition of localized and nonlocalized surface diffusion, the various diffusing defects will have a different distribution of degrees of freedom in each state. This is also indicated in Fig. 3b in the case of an adatom. More complicated distributions are valid for higher order molecular defects, such as dimers, trimers, etc. (Bonzel, 1970).

In the following we will calculate intrinsic diffusion coefficients for localized as well as nonlocalized states. When activated states are involved the superscript $^{+}$ will generally be used; however, in the specific case of localized diffusion we will use the superscript ‡ and for nonlocalized diffusion the superscript *.

If a defect makes random jumps of average length l, where τ is the average time between successive jumps, the intrinsic diffusion coefficient is given by random walk theory (see e.g., Feller, 1959) as

$$D = \alpha(l^{2}/\tau), \tag{10}$$

with α as a numerical constant ($\frac{1}{2}$ for one-dimensional motion, $\frac{1}{4}$ for surface diffusion, $\frac{1}{6}$ for bulk diffusion, in a simple cubic system). We now introduce an average velocity of a defect $\bar{V}$ while in the diffusive state and a time τ^{+}, which is the average time of activation such that $l = \bar{V}\tau^{+}$. Equation (10) then becomes

$$D = \alpha \bar{V}^{2}\tau(\tau^{+}/\tau)^{2}. \tag{11}$$

This equation was used by Lennard–Jones (1932, 1937) as a starting point for calculating surface diffusion coefficients. He assumed that vibrational and mobile states, defined by τ and τ^+ respectively, are characterized by discrete energy levels. If we make use of the theorem that for large thermodynamic systems time averages are equal to spacial averages (e.g., Hill, 1960), we may replace the ratio τ^+/τ by a concentration ratio C^+/C, where C^+ and C are the concentrations of defects in the mobile and vibrational state respectively. These are then related to their respective partition functions, F^+ and F:

$$(\tau^+/\tau) = (C^+/C) = (F^+/F) \exp(-\Delta H^+/kT), \tag{12}$$

where ΔH^+ is the energy of transfer from vibrational into mobile state (energy of migration).

The relaxation time τ in Eq. (10) can be obtained from Frenkel's (1924) expression, which was also discussed in detail by deBoer (1968),

$$\tau = \tau_0 \exp(\Delta H^+/kT), \tag{13}$$

where τ_0 is $1/v$, with v as the basic frequency along the reaction coordinate.
For localized adatom diffusion we choose

$$\bar{V} = \left(\frac{kT}{2\pi m}\right)^{1/2}, \tag{14}$$

$$\frac{F^\ddagger}{F} = \left(\frac{2\pi m k T}{k^2}\right)^{1/2} l \frac{hv}{kT}, \qquad T \gg \frac{hv}{k}, \tag{15}$$

where k is Boltzmann's constant, m is the atomic mass, and k is Planck's constant.

By inserting Eqs. (12) to (15) into Eq. (11) we obtain

$$D = \alpha l^2 v \exp(-\Delta H^\ddagger/kT). \tag{16}$$

A minor entropy term in Eq. (13) has been neglected, and therefore no entropy term appears in Eq. (16) for localized motion.

Equation (16) was derived for the case of adatom diffusion where $\Delta H^\ddagger$ is the energy of migration. A completely analogous expression can be derived for vacancy migration, and furthermore quite similar expressions hold for molecular surface diffusion as long as the diffusing defect in its mobile state exhibits only one degree of translational freedom. However, the formalism is somewhat different if excitations into the nonlocalized state occur. In the case of an adatom Fig. 3b shows that the nonlocalized state is characterized by two translational and one vibrational degree of freedom. For a dimer the corresponding distribution would call for two translational, two rotational, and two vibrational degrees of freedom (two-dimensional gas) or, if an intermediate case exists, two translational, one rotational, and three vibrational degrees of freedom. A fully excited trimer would exhibit two translational, three rotational, and four vibrational degrees of freedom.

In general the partition functions F^* of a defect in the nonlocalized state are a product of translational, rotational, and vibrational partition functions, where the latter are assumed to be equal to those of free gas molecules (in two dimensions). The vibrational partition function is that of a harmonic oscillator. The intrinsic diffusivity of a dimer in the nonlocalized state is then, for example,

$$D_2^* = \alpha \frac{2^{13}\pi^5 m}{v_6 kT} \left(\frac{A I v_2 v_3 v_4 v_5}{kT} \right)^2 \exp\left(-\frac{\Delta H_2^*}{kT} \right), \tag{17}$$

where A is the area per surface site; I is the moment of inertia of the dimer; $v_2, v_3, \ldots, v_6$ are the vibrational frequencies of the dimer in vibrational state; ΔH_2^* is the energy to transfer the dimer from localized to nonlocalized state. If we write Eq. (17) in the general form

$$D_2^* = \alpha l^2 v_6 \exp\left(\frac{\Delta S_2^*}{k} \right) \exp\left(-\frac{\Delta H_2^*}{kT} \right), \tag{18}$$

we see that the entropy of migration of a dimer in the nonlocalized state ΔS_2^* can be calculated from

$$\exp\left(\frac{\Delta S_2^*}{k} \right) = \frac{2^{13}\pi^5 m A}{kT} \left(\frac{I v_2 v_3 v_4 v_5}{v_6 kT} \right)^2. \tag{19}$$

It has been shown for the diffusion of the dimers that this entropy factor can be very large compared to those expected for localized diffusion (Bonzel, 1970). Furthermore the deactivation of defects from the nonlocalized state may be difficult because of a resulting change in bond length of dimer and trimer. Therefore the transition from nonlocalized to localized state may be characterized by an activation energy (Hirth, 1970).

Surface diffusion in the nonlocalized state leads to long jump distances of diffusing defects, which follows from the expression

$$l^* = \bar{V}^* \tau^* = \bar{V}^* \tau_0 F^*/F. \tag{20}$$

The ratio of partition functions in Eq. (20) will be very large because of the introduction of translational and rotational degrees of freedom into F^*. Earlier Shewmon and Choi (1962) proposed long jump distances for adatoms in order to explain large preexponential factors of measured diffusion coefficients. For a more detailed discussion of their approach one should consult the review by Bonzel (1973).

V. PHENOMENOLOGY OF SURFACE DIFFUSION

Surface diffusion coefficients are related to measurable parameters by means of phenomenological equations. These are generally derived for a given geometry and boundary conditions. According to the various experimentally suitable geometric arrangements, we distinguish four groups of phenomenological equations discussed in the following sections.

A. Random Walk

The definition of a diffusion coefficient by random walk theory, Eq. (10), has been used to measure surface diffusion coefficients. Such measurements can be carried out in the absence as well as the presence of concentration gradients and for both self-diffusion and heterodiffusion. In order to make use of Eq. (10) for measuring D, one in essence observes one atom or many atoms (in a diffusion front) over a large number of jumps, say n. For these n jumps the mean square displacement $\bar{x}^2$ of the moving atoms and the total time t are measured. Then, because the mean time of stay τ is equal to t/n, one has

$$\frac{\sum_{i=1}^{n} l_i^2}{\sum_{i=1}^{n} \tau_i} = \frac{nl^2}{t} = \frac{\bar{x}^2}{t} = \frac{l^2}{\tau}, \tag{21}$$

and therefore

$$D = \alpha \frac{\bar{x}^2}{t}. \tag{22}$$

Ehrlich and Hudda (1966) demonstrated how the migration of individual tungsten adatoms could be followed by field ion microscopy. In these experiments the motion of adatoms was restricted to flat terraces, and only an occasional interaction with a ledge was observed. Under those conditions Eq. (22) may be applied, but for smaller terraces and frequent boundary (ledge) reflections a statistically more relevant expression must be used (Ehrlich, 1966).

On the other hand, if a large number of atoms diffuse, such as in an advancing boundary representative of a steep concentration gradient (Gomer, 1959), Eq. (22) can also be used to measure D. This is true as long as the concentration distribution $c(x, t)$ may be written in the form $c(x^2/4Dt, t)$. This is particularly easy to see for the so-called infinite source equation, because

$$\bar{x}^2 = \frac{\int_0^\infty x^2 c(x,t)\,dx}{\int_0^\infty c(x,t)\,dx} = 2Dt, \tag{23}$$

$$c(x,t) = \frac{s}{2(\pi Dt)^{1/2}} \exp\left(-\frac{x^2}{4Dt}\right), \tag{23a}$$

where s = concentration of source. In this case $(\bar{x}^2)^{1/2}$ is set equal to the linear advance of a sharp boundary, an experimental signal of the concentration gradient.

B. Surface Diffusion of Distinguishable Atoms on a Flat Surface

Quite generally surface diffusion takes place in a chemical potential gradient field and is observed as changes in concentration of the diffusing species or changes in topography of the sample surface. If $\mu(x)$ is the chemical potential in one dimension, the drift velocity v_i of a defect "i" is given by the Nernst–Einstein relation (Jost, 1952)

$$v_i = -\frac{D_i}{kT}\frac{\partial \mu}{\partial x}. \tag{24}$$

With the concentration n_i of the diffusing defect the diffusion flux J_i is equal to:

$$J_i = n_i v_i = -\frac{n_i D_i}{kT}\frac{\partial \mu}{\partial x}. \tag{25}$$

Here we define a mass transfer diffusion coefficient by the equation

$$D_s = (n_i/N_0)D_i \tag{26}$$

where N_0 is the total number of adsorption sites per unit area, $N_0 = \Omega^{-2/3}, \Omega$ = atomic volume. The fractional concentration, $c_i = n_i/N_0$, as a function of temperature is given by (Burton $et\ al.$, 1951; Leamy $et\ al.$, Chapter 3 of Volume I)

$$c_i = c_{i0} \exp(-\Delta G_f^i/kT), \qquad c_{i0} = \text{constant} \tag{27}$$

where ΔG_f^i is the free energy of formation of the defect i. If pairs or higher order defects can be formed by the defect i, the concentration of pairs, for example, is

$$c_{pi} = c_{i0}^2 \exp\left(-\frac{2\Delta G_f^i - B_i}{kT}\right), \tag{27a}$$

where B_i is the binding energy of the pair.

If several diffusing defects can contribute to the overall diffusion flux, and, moreover, if more than one mode of surface diffusion is possible, the total flux can be written as the sum of all individual fluxes. Hence

$$D_s = \sum_{i=1}^{n} c_i (D_i^{\ddagger} + D_i^*) \tag{28}$$

for localized and nonlocalized surface diffusion, where c_i are fractional surface concentrations of the diffusing defects.

However, one should remember that these equations are only valid as long as local thermodynamic equilibrium can be maintained (with respect to defect concentrations). This is particularly important in the case of tracer diffusion. For example, if $n'(x)$ is the concentration profile of a radioactive tracer atom on the surface, then the chemical potential is given by

$$\mu(x) = \mu(0) + kT \ln \frac{n'(x)}{N_0}. \tag{29}$$

Because of the complex defect structure of a crystalline surface, $n'(x)$ is generally not equal to $n_i'(x)$, the concentration of the species in the diffusive state (e.g., adatom). Many tracer atoms can be located at kinks, ledges, etc. Calculating $\partial\mu/\partial x$ and inserting it into Eq. (26) yields, as discussed by Gjostein and Hirth (1965) and Blakely (1966),

$$J_i = -\frac{n_i' D_i}{n'} \frac{\partial n'}{\partial x}. \tag{30}$$

We can now distinguish two different cases:

(a) If ledges, kinks, etc. were totally absent, that is, for an ideal terrace $n_i' = n'$, then the measured diffusivity would be D_i, an intrinsic diffusivity. On a more realistic surface the same would hold only if all surface defects were blocked such that tracer atoms could not adsorb at these sites. This may be expected for a highly contaminated surface.

(b) On a clean TLK surface the tracer atoms are likely to equilibrate with all defects. For such an equilibrium a relationship of the following type might exist:

$$n_i'/n' = A \exp(-\Delta H_f/kT), \tag{31}$$

where ΔH_f is a formation energy for the diffusive defect, e.g., an adatom. For a fully equilibrated surface this energy would be identical to the formation energy of an adatom from a kink site, as outlined previously. Then

$$J_i = -AD_i \exp(-\Delta H_f/kT)(\partial n'/\partial x). \tag{32}$$

The measured diffusivity is therefore a mass transfer diffusivity D_s.

In this context it is important to realize that under certain experimental conditions a finite time t_1 is needed to approach equilibrium between surface defects and the diffusing adatom concentration. This relaxation time t_1 can be compared to the actual duration of an experiment t_e, and if $t_1 \gtrsim t_e$, Eq. (32) does not hold. It is conceivable that during vapor deposition experiments, for instance, a high nonequilibrium concentration of diffusive defects may be present on the surface because of the associated high supersaturation of adatoms. In such a case any observed diffusivity would be closer to an intrinsic diffusivity.

In addition to the problem of assessing the degree of thermodynamic equilibrium during a tracer diffusion experiment, there are some other problems with this approach of measuring surface diffusion. These are: (1) influence of impurities, (2) effect of bulk diffusion and evaporation of tracer, (3) exact mathematical solutions for the diffusion process, (4) experimental difficulties with obtaining correct concentration profiles at various times.

There are three different types of geometric sources that are, from an experimental point of view, attractive and that have been treated mathematically. Figure 4 shows schematically an edge source, half-plane source, and point source of tracer material. Several approximate mathematical solutions have been reported by Geguzin *et al.* (1969), Drew and Pye (1963), Shewmon (1963), and Drew and Amar (1964); these are all more or less

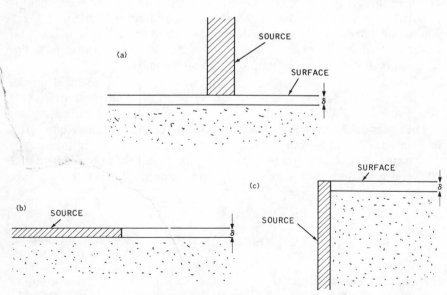

Fig. 4. Schematic representation of geometric sources in tracer surface diffusion experiments. (a) Point source; (b) Half-plane source; (c) Edge source.

based on early treatments of grain boundary diffusion by Fisher (1951) and Whipple (1954). However, a rigorous treatment of tracer surface diffusion has been given by Suzuoka (1965) for edge and half-plane sources.

A recent summary of formulas and a discussion of the evaluation of experimental results was given by Gjostein (1970a).

C. Surface Diffusion Due to Capillarity Forces

A driving force for surface diffusion different from a concentration gradient of species A on substrate B is due to capillarity, or the minimization of the surface free energy. The basic equation relating the shape of a surface in terms of a principal curvature $K(x)$ to the chemical potential $\mu(x)$ is the classical Gibbs–Thomson formula (Herring, 1951):

$$\mu(x) = \left(\gamma(\theta) + \frac{\partial^2 \gamma}{\partial \theta^2} \right) \Omega K(x), \qquad (33)$$

where $\gamma(\theta)$ is the orientation dependent surface energy and

$$K(x) = -\frac{d^2 y}{dx^2} \left(1 + \left(\frac{dy}{dx} \right)^2 \right)^{-3/2}, \qquad (34)$$

$y(x)$ describing a surface profile. Equation (33) is valid only for such orientations where $\gamma'' = d^2\gamma/d\theta^2$ exists (Mullins, 1963). If surface diffusion takes place over a small range of crystallographic orientations, γ and γ'' may be replaced by average values γ_0 and γ_0''. In this case the diffusion flux, Eq. (25), can be obtained for capillarity driven diffusion as

$$J = -\frac{(\gamma_0 + \gamma_0'') N_0 D_s \Omega}{kT} \frac{\partial K}{\partial x}. \qquad (35)$$

The diffusion flux causes the surface profile to change. The rate of motion of a surface element normal to the surface is obtained by taking the surface divergence of $-J$ and multiplying it by Ω. The result is a differential equation governing the rate of change of a surface profile $y(x)$:

$$\frac{dy}{dt} = -B \frac{d^4 y}{dx^4}, \qquad \frac{dy}{dx} \ll 1 \qquad (36)$$

$$B = \frac{(\gamma_0 + \gamma_0'') N_0 \Omega^2 D_s}{kT}. \qquad (37)$$

It should be noted that surface diffusion due to capillarity forces is generally limited to self-diffusion, with the exception of particle coalescence (Ostwald ripening).

Equation (36) can now be solved for a variety of $y(x)$ functions and boundary conditions. There is a growing list of geometric shapes for which solutions are known:

(a) *Grain boundary groove.*   An initially flat polished surface develops a groove at a grain boundary, since surface energy and grain boundary energy want to be equilibrated as shown by Bailey and Watkins (1950). The solution for surface diffusion controlled grain boundary grooving was derived by Mullins (1957).

(b) *Periodic surface profile (sinusoidal profile).*   A deliberately generated periodic surface profile, such as a set of parallel scratches or etched grooves, will relax to flatness under the action of surface diffusion, bulk diffusion, evaporation–condensation (closed system), and viscous flow. The solution to this problem was given by Mullins (1959) and has been applied many times to the measurement of surface self-diffusion of metals and oxides.

(c) *Nonperiodic surface profile.*   The simplest of these profiles is that of an isolated scratch in an originally flat surface. The theory of the decay of such a scratch was developed by King and Mullins (1962).

(d) *Tip blunting.*   Field electron emitter tips used in field emission microscopy show blunting at elevated temperature. This process is capillarity driven and governed by surface diffusion, as shown by Nichols and Mullins (1965a).

(e) *Sintering and void migration.*   The behavior of an ensemble of spherical particles or voids is quite similar and generally governed by surface diffusion. A discussion of these processes and solutions to the problem were given by Nichols and Mullins (1965b), Nichols (1966), Barnes and Mazey (1963), and Shewmon (1964).

(f) *Linear faceting.*   Under certain conditions an initially flat surface develops facets that have a discretely different orientation from the neighboring surface. The lateral growth of facets can also be used to evaluate surface diffusion coefficients, as shown by Mullins (1961).

(g) *Particle coalescence (Ostwald ripening).*   An ensemble of small particles on a flat substrate surface may rearrange at elevated temperature in order to lower the total surface and interfacial free energy of the system. Large particles will grow at the expense of small particles. The mechanism is thought to be surface diffusion of adatoms from higher to lower chemical potential (i.e., from small to large particles). This problem has been treated mathematically by Chakraverty (1967), Geguzin *et al.* (1969), and more recently Abramenkov *et al.* (1971), Other relevant work in this area was reported by Lifshitz and Slyozov (1961) and Wagner (1961). It is noteworthy that in these experiments a heterodiffusion rather than self-diffusion coefficient will be measured in contrast to all other techniques described in this section.

A particularly simple, but nevertheless very useful, example of a solution of Eq. (36) is a periodic function of $y(x)$ as described under (b). For

$$y(x,t) = \sum_{n=-\infty}^{\infty} A_n(t) \exp(in\omega x) \qquad (38)$$

with $\omega = 2\pi/\lambda$, λ being the fundamental wavelength of the profile, and $A_n(t)$ as the time dependent amplitude coefficients, Mullins (1963) gave the solution

$$A_n(t) = A_n(0) \exp(-Bn^4\omega^4 t), \qquad (39)$$

$$A_n(0) = 1/2\pi \int_0^\lambda y(x,0) \exp(-in\omega x)\,dx. \qquad (40)$$

In the special case $y(x,0) = A(0)\sin(\omega x)$, the solution simplifies to

$$A(t) = A(0) \exp(-B\omega^4 t). \qquad (41)$$

From Eq. (39) it is easy to see that any arbitrary periodic profile will finally become sinusoidal during the smoothing process, because all higher harmonics $n\omega$ $(n > 1)$ will decay with a decay constant n^4 times faster than that of the fundamental wave ω.

Besides surface diffusion there are generally other transport processes contributing to the morphological changes under discussion. This may be particularly true at higher temperatures. For the case of a sinusoidal profile it is relatively simple to include contributions due to volume diffusion, diffusion through the gaseous phase, evaporation–condensation, and viscous flow (Mullins, 1959). In such a more general case the equation for the time dependence of the amplitude is

$$A(t) = A(0) \exp[-(F\omega + A\omega^2 + D\omega^3 + B\omega^4)t], \qquad (42)$$

$$\text{viscous flow:} \quad F = \gamma/2v, \qquad (43)$$

$$\text{evaporation–condensation:} \quad A = \frac{p_0 \gamma \Omega^2}{(2\pi M)^{1/2}(kT)^{3/2}} \qquad (44)$$

$$\text{volume diffusion:} \quad D = A' + C$$

$$A' = \frac{\rho_0 D_G \gamma \Omega^2}{kT}, \qquad C = \frac{D_v \gamma \Omega}{kT}, \qquad (45)$$

where v is the viscosity of the solid, p_0 is the vapor pressure of the solid, M is the mass of the molecule, ρ_0 is the vapor density over a surface of zero curvature, D_G is the diffusion coefficient of vapor molecules in gas phase, D_v is the volume diffusion coefficient of the solid. By knowing most of these quantities one should select the experimental conditions (for a measurement of surface diffusion) such that $F\omega + A\omega^3 + A'\omega^3 \ll C\omega^3 + B\omega^4$, or even at lower temperature, that $B\omega^4$ is the dominant term. Such consideration can largely simplify any experiment.

As pointed out earlier, all phenomenological equations governing morphological shape changes due to capillarity are derived on the basis of isotropic surface energy and diffusion coefficient. There are indications from several experiments that this limitation of isotropic γ is not always compatible with actual observations. A method for including the anisotropy of $\gamma(\theta)$ in capillarity theory is based on perturbation analysis of $\gamma(\theta)$ and the shape function of the surface. This method was first used by Cahn (1967) in treating shape changes of precipitate particles by volume diffusion and has recently been applied by Gjostein and Brailsford (1973) to perturbed cylinders and spheres where shape changes occur by surface diffusion. We will briefly outline the method for the example of an infinite cylinder. The problem is solved for the following assumptions:

1. The chemical potential at any point P of the surface is given as a function of the two principal curvatures K_1 and K_2 and of the orientation dependent $\gamma(\theta)$ in the following way:

$$\mu(P) = \mu(0) + \Omega \left[\left(\gamma + \frac{\partial^2 \gamma}{\partial \theta_1^{\,2}} \right) K_1 + \left(\gamma + \frac{\partial^2 \gamma}{\partial \theta_2^{\,2}} \right) K_2 \right] \qquad (46)$$

2. The surface energy $\gamma(\theta)$ for cylindrical geometry is expressed as

$$\gamma(\theta) = \gamma_0 \left[1 + \sum_m a_m \cos(m\theta) \right], \qquad (47)$$

where the a_m are constants.

3. The shape of the surface is described by a radius vector R,

$$R(\theta) = R_0 \left[1 + \sum_m \delta_m(t) \cos(m\theta) \right]. \qquad (48)$$

4. For $a_m, \delta_m \ll 1$, it is assumed that for a point P, defined by the vector $R(\theta)$, the orientation angle θ' is actually equal to θ (difference between tangent at P and normal to R).

5. The surface diffusion flux at P is as follows:

$$J(P) = -\frac{N_0 D_s}{kT} \text{grad}_s \left[\mu(P) - \mu(0) \right]. \qquad (49)$$

From the form of Eq. (49) it is obvious that the anisotropy of D_s is neglected. It is also important to note that the surface gradient operator grad_s differentiates $\gamma(\theta)$ and $\gamma''(\theta)$ as well as K.

By neglecting cross terms of the type $\delta_m a_m$ a simple solution can be obtained.

$$\delta_m(t) = \delta_m^{eq} + [\delta_m^{eq} - \delta_m(0)] \exp\left\{- m^2(m^2-1)\frac{B}{R_0^4}t\right\} \qquad (50)$$

$$\delta_m^{eq} = a_m, \qquad (51)$$

where B is given by Eq. (37). Thus the equilibrium shape in this analysis is geometrically similar to the γ-plot. The particle becomes flatter for such orientations corresponding to dimples in the γ-plot; note that all orientations are present on the particle shape.

The presence of sharp corners (missing orientations) on the shape plot is not seen in this simple, linear analysis; but if cross terms of the type $a_m \partial_m$ are included, morphological instabilities occur wherever $\gamma + \gamma'' < 0$. These instabilities reflect the fact that the Fourier interval employed in the perturbation analysis must be redefined to avoid the discontinuities represented by the sharp corners.

Gjostein and Brailsford (1973) also treat the more complicated case of a sphere where the shape function and γ-plot depend on two independent variables. This case should be particularly important for the coalescence of supported particles since those particles exhibit a large range of crystallographic orientations and are frequently faceted (Wynblatt and Gjostein, 1974). If extensive faceting occurs, the growth law may be substantially different from that derived for the isotropic case (Chakraverty, 1967).

D. Surface Diffusion during Whisker Growth

Volmer and Estermann (1921) observed the growth of thin mercury platelets from the supersaturated vapor and interpreted the anisotropy of growth by a three-stage mechanism:

(a) deposition of atoms on all surfaces of the platelet;
(b) rapid surface diffusion of adatoms to the periphery of the platelet;
(c) incorporation of adatoms into crystal lattice at the edges of the platelet.

Sears (1951, 1953) reinvestigated the growth of mercury crystals from the vapor phase and found that not only platelets but also thin whiskers ($\sim 10^{-5}$ cm radius) could be grown under certain conditions. The one-directional growth was explained by the existence of screw dislocations with their Burgers vectors parallel to the growth direction. The kinetics of whisker growth seemed to obey the general picture outlined above, namely that adatoms deposited from the vapor phase would diffuse along the surface of the whisker until they either evaporated or became incorporated into the lattice at the tip of the whisker causing it to grow longer. Further studies were carried out by Sears (1955, 1956) and by Dittmar and Neumann (1960), who also derived quantitative growth laws. Other derivations of growth laws were reported by Gomer (1958), Blakely and Jackson (1962), Hirth and Pound (1963), and Ruth and Hirth (1964). Particularly this last paper gives a detailed discussion of the validity ranges of various steady state growth laws and points out mechanisms for cessation of growth or the transition from exponential to linear growth.

Since whisker growth involves the surface diffusion of adatoms to the tip of the whisker (and the base of the whisker as well as the substrate itself), it is clear that a measurement of the growth kinetics should yield information on the rate of surface diffusion. The general relationship expected to exist between the rate of whisker growth and the surface diffusion coefficient is derived from the continuity equation subject to certain boundary conditions (Ruth and Hirth, 1964):

$$\frac{\partial c(x',t)}{\partial t} = D \frac{\partial^2 c(x',t)}{\partial x'^2} - \omega c(x',t) + F. \tag{52}$$

In Eq. (52) D is the surface diffusion coefficient of adatoms, $c(x',t)$ the concentration of adatoms along the whisker surface, ω the probability for desorption of adatoms, and F the impingement flux from the vapor. It is advantageous to use a moving frame of reference, with the tip of the whisker at the origin $x = 0$. Then one obtains

$$\frac{\partial c(x,t)}{\partial t} = D \frac{\partial^2 c(x,t)}{\partial x^2} - V(t) \frac{\partial c(x,t)}{\partial x} - \omega c(x,t) + F, \tag{53}$$

where

$$V(t) = \frac{dL}{dt} = \alpha D \left. \frac{\partial c(x,t)}{\partial x} \right|_{x=0} + F' \tag{54}$$

is the growth rate of the whisker, L is the length of the whisker, F' is the impingement flux onto the whisker tip.

Simple solutions to Eq. (53) are obtained when steady state conditions are allowed such that

$$\frac{\partial c(x,t)}{\partial t} = 0, \qquad \frac{\partial V(t)}{\partial t} = 0. \tag{55}$$

We choose as boundary conditions (see Fig. 5)

$$c(0,t) = c_0, \qquad c(\infty,t) = c_\infty \tag{56}$$

where $x = 0$ is at the tip of the whisker and $x \to \infty$ designates the base

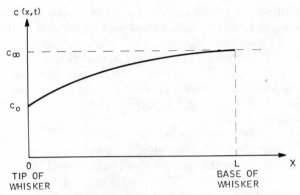

Fig. 5. Adatom concentration profile on the surface (shank) of a whisker. c_0 is the equilibrated adatom concentration at the tip of the whisker, representative of a temperature T, while c_∞ is a supersaturated adatom concentration at the base depending on the gas pressure.

(asymptotic case, concentration gradient at the base is zero). The concentration c_0 is in thermodynamic equilibrium with the vapor. These conditions lead to a linear growth law as shown by Ruth and Hirth (1964):

$$\frac{dL}{dt} = \frac{2\Omega}{R}(D\omega)^{1/2}(c_\infty - c_0), \tag{57}$$

with R as the radius of the whisker at the tip. It can be seen in Eq. (57) that quite a number of quantities have to be known or measured in order to permit a measurement of D. First of all the rate of whisker growth, dL/dt, will have to be measured directly. The concentration c_∞ is usually determined by equating the impingement flux F and the reevaporation flux $c_\infty \omega$ such that

$$c_\infty = F/\omega. \tag{58}$$

Therefore the quantities c_0 and ω should be known independently in order to calculate D. As pointed out by Dittmar and Neumann (1960), the integration of Eq. (57) can be used to determine $D\omega$ and R simultaneously. However, for a more detailed discussion of these specific points the reader is referred to the original literature.

An exponential growth law can be obtained by altering the boundary conditions, Eq. (56), namely

$$c(0, t) = c_0; \left.\frac{\partial c(x, t)}{\partial x}\right|_{x=L} = 0. \tag{59}$$

The latter condition in Eq. (59) means that the gradient of adatoms at the base of the whisker is always zero. The growth law is then

$$\frac{dL}{dt} = \frac{\sqrt{2}\bar{x}\Omega\omega}{R}(c_\infty - c_0)\tanh\left(\frac{\sqrt{2}L}{\bar{x}}\right) + F', \tag{60}$$

where

$$\bar{x} = (2D/\omega)^{1/2} \tag{61}$$

is the mean root square diffusion distance of an adatom on the whisker surface. Equation (60) reduces in the limit $h > \bar{x}$ to the linear law, Eq. (57), while in the limit $h < \bar{x}$ and negligible F' an exponential law obtains

$$\frac{dL}{dt} = \frac{2\Omega\omega L}{R}(c_\infty - c_0). \tag{62}$$

It is often observed experimentally that whiskers grow exponentially at first [or according to Eq. (60)] and that at some stage exponential growth changes into linear growth or ceases completely. Ruth and Hirth (1964) have pointed out a number of reasons for this transition, specifically, that screw dislocations responsible for whisker growth may slip or climb out of the whisker, or that impurities at the whisker tip may decrease the rate of growth. Reevaporation of adatoms and a violation of the tip boundary condition can also change the rate law, and even two-dimensional nucleation on the whisker sides could sufficiently decrease the concentration of adatoms to lower the rate of growth. On the other hand, a violation of the steady state condition should never be the reason for a change of rate law.

It should be noted that the D measured by this technique is always a self-diffusion coefficient. Moreover, D is most likely an intrinsic diffusion coefficient characterized by the energy of adatom migration. The mere fact that it is possible to grow whiskers from the vapor phase implies that the adatom population on the whisker sides is not equilibrated with surface defects, because the only place where equilibration takes place is at the tip of the whisker. Therefore one may conclude that the density of steps and kink sites on the whisker sides is negligible.

VI. EXPERIMENTAL TECHNIQUES

A detailed description of the available experimental techniques for measuring surface diffusion coefficients on crystalline surfaces will not be presented in this article since it would take up too much space. The interested reader should consult other review papers listed in Table I or suitable original papers.

TABLE IV

SURVEY OF EXPERIMENTAL TECHNIQUES (FOR MEASURING SURFACE DIFFUSION)

Classification of measuring techniques	Measured diffusion coefficient			References
	SD	HD	I	
A Field ion emission microscopy (FIM)				
A1 Adsorbed atoms, random walk	X	X	X	a
B Field electron emission microscopy (FEM)				
B1 Ring rate technique, tip blunting	X			b
B2 Field build-up protrusion decay	X			c
B3 Spreading of adsorbate		X		d
C Bulk sample, capillarity driven				
C1 Grain boundary grooving	X			e
C2 Sinusoidal profile decay	X			f
C3 Single scratch decay	X			g
C4 Linear facet growth	X			h
C5 Sintering of small particle	X			i
C6 Void migration	X			j
C7 Coalescence of particles		X		k
D Bulk sample, concentration gradient				
D1 Radioactive tracer	X	X	?	l
D2 Change in work function, thermoionic emission due to adsorbate		X		m
D3 Secondary ion emission		X		n
D4 Auger electron spectroscopy		X		o
E Vapor deposition (concentration gradient)				
E1 Growth of whiskers and platelets	X		X	p
E2 Nucleation and growth of small particles on a substrate		X	?	q
E3 Diffusion during vapor deposition		X		r

[a] Müller (1957); Ehrlich and Hudda (1966).
[b] Müller (1949); Dyke and Barbour (1956); Barbour et al. (1960).
[c] Sokolskaya (1956); Bettler and Charbonnier (1960).
[d] Drechsler (1954); Gomer (1959); Melmed (1965); Vladimirov et al. (1970).
[e] Mullins and Shewmon (1959); Gjostein (1961).
[f] Blakely and Mykura (1962); Gjostein (1965); Bonzel and Gjostein (1969).
[g] Blakely and Mykura (1961); Shewmon and Choi (1963).
[h] Rhead and Mykura (1962).
[i] Kuczynski (1959, 1961).
[j] Barnes and Mazey (1963); Shewmon (1964); Willertz and Shewmon (1970).
[k] Geguzin et al. (1969); Kaganovskiy and Ratinov (1971); Wynblatt and Gjostein (1974).
[l] Nickerson and Parker (1950); Wolfe and Weart (1969).
[m] Langmuir and Taylor (1932); Barrer (1941).
[n] Abramenkov et al. (1971).
[o] Bonzel and Ku (1973).
[p] Sears (1953, 1955); Dittmar and Neumann (1960).
[q] Sumner (1965); Trofinov et al. (1969).
[r] Roulet (1973).

In Table IV the experimental techniques are summarized, and it is indicated whether a self-diffusion (SD) or a heterodiffusion (HD) coefficient can be measured by the technique and also whether or not the measured diffusion coefficient is an intrinsic (I) diffusivity, i.e., independent of the energy of formation of a defect. The techniques are classified somewhat arbitrarily according to the scale of observation (microscopic versus macroscopic) and driving force (capillarity versus concentration gradient).

It is quite apparent from Table IV that most experimental techniques have been developed for measuring SD coefficients of the mass transfer type, where the activation energy is the sum of formation and migration energy of defects. Very few techniques are capable of measuring an intrinsic diffusion coefficient reliably. The intrinsic diffusivities determined from whisker growth experiments still suffer from the difficulty of separating the product Dc_∞ (cf. Section V.D).

The great advantage of the experiments in groups A and B of Table IV is the fact that they are carried out in extremely good vacuum of $\leqslant 1 \times 10^{-10}$ Torr and under very good cleanliness conditions. Therefore a measured diffusion coefficient will most likely be representative of an atomically clean surface. In addition the scale factor (Gjostein, 1973) is so small that diffusion can be observed at such low temperatures that impurity segregation effects are less likely to occur because of low bulk diffusion rates.

Another technique that permits one to establish, monitor, and control the cleanliness of the surface on which surface diffusion takes place is the sinusoidal profile decay technique carried out in situ (Bonzel and Gjostein, 1967). The intensity distribution of a laser diffraction pattern generated by the periodic (sinusoidal) surface profile allows a measurement of the amplitude of the sine wave and thus a continuous measurement of the surface diffusion rate at constant temperature. At the same time surface analytical techniques such as LEED and Auger electron spectroscopy can be used to monitor surface cleanliness (Bonzel and Gjostein, 1968; Blakely and Olson, 1968; Bonzel, 1973).

On the other hand many techniques do not easily permit a cleanliness control of the surface or even a characterization of the surface, and hence the results obtained are quite often suspect because of possible impurity effects. The latter have been demonstrated to exist in a number of cases (Bonzel and Gjostein, 1968a, 1969; Olson et al., 1972; Perdereau and Rhead, 1967), and it is interesting to note that a suppression as well as an enhancement of the surface diffusion rate has been observed depending on the kind of impurity present (Bonzel, 1973). This is exactly the reason why the interpretation of experimental surface diffusion results and a critical comparison of experiment and theory is very difficult. Only in a few cases is it possible to state unambiguously whether or not surface diffusion representative of a clean surface was measured.

VII. EXPERIMENTAL RESULTS AND DISCUSSION

In the following sections we will present some experimental results of surface diffusion measurements that will be used for a general comparison of theory and experiment and also for a comparison of experimental data alone, since they were obtained by different techniques. Most of the experiments were carried out for metals (one-component systems), self-diffusion as well as heterodiffusion. A few experiments dealing with surface diffusion on alloy surfaces have also been published, and a few for oxides, carbides, and alkali halides are known. We will present the results according to this classification of materials. Finally we will skim the surface of such diverse subjects as coalescence of particles, Brownian motion of particles, and diffusion on porous media surfaces.

A. Surface Self-Diffusion of Metals

As pointed out in Section V, surface self-diffusion coefficients fall into two categories: mass transfer and intrinsic diffusion coefficients. The latter are much more difficult to measure, but both quantities are equally important for an understanding of the behavior of surface defects. If the temperature dependence of both diffusion coefficients for a given material is known, the formation *and* migration energy of the diffusing defect can be determined. The separation of these two energies is here just as difficult and important as in the field of bulk lattice diffusion.

We will first discuss measurements of mass transfer diffusion coefficients. There are three important general results that have crystallized over the past ten years and should receive special attention. The first is the excellent agreement between phenomenological theory of capillarity induced morphological changes and experimental kinetics. For example, in the case of grain boundary grooving the width of the groove should increase proportional to $t^{1/4}$ (Mullins, 1957); this is in fact the case, as shown in Fig. 6, if the bulk diffusion contribution is properly subtracted (Gjostein, 1970b). Even though this particular issue has been the subject of controversy (McAllister and Cutler, 1970, 1972), it has been shown convincingly that indeed surface and bulk diffusion contribute to the growth of a grain boundary groove (Srinivasan and Trivedi, 1973).

Similarly to Fig. 6, Fig. 7 shows a plot of log ($2A$) versus time in accordance with Eq. (41) for the decay of a sinusoidal surface profile (Bonzel and Gjostein, 1968a).[†] Straight lines are obtained for various periods of the profile

[†] $2A$ is the depth of the profile.

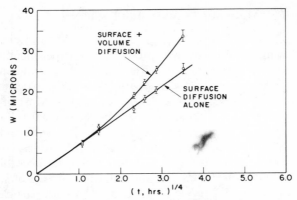

Fig. 6. Plot of the width of a grain boundary groove versus annealing time for a copper crystal at $T = 1020°$ C; note that the points due to surface diffusion alone fall on a straight line that goes through the origin (Gjostein, 1970b).

and at different temperatures. Furthermore, in the case of simultaneous bulk and surface diffusion, the slope of a plot of $\log(2A)$ versus t, $S(\lambda, T)$ should be a function of the constants D and B of Eq. (42), if viscous flow and evaporation can be neglected.

This means that

$$S(\lambda, T) = -\log e(D\omega^3 + B\omega^4), \tag{63}$$

or with $A' \approx 0$,

$$\lambda^4 S(\lambda, T) = -\frac{\log e}{(2\pi)^4}\left(\frac{C}{2\pi}\lambda + B\right). \tag{64}$$

Therefore, if surface diffusion measurements are carried out for several values of λ at constant temperature, a plot of $\lambda^4 S(\lambda, T)$ versus λ should yield straight lines whose slopes are proportional to the bulk diffusion coefficient and whose intercepts with the ordinate are proportional to the surface diffusion coefficient. Such measurements were performed for Cu(111) crystals by Hoehne and Sizman (1971). Their data showed excellent agreement with the dependence of Eq. (64), but their final evaluation suffered from an error (see discussion by Bonzel and Gjostein, 1972). Also Maiya and Blakely (1967) had earlier investigated surface self-diffusion of nickel by the same technique for $\lambda = 6\mu$ to $\lambda = 38\mu$ and found that a plot of $\log S(\lambda, T)$ versus $\log \lambda$ for constant temperature and after subtraction of $C\omega^3$ was characterized by a slope of -4.

Although we have cited only a few investigations, there are numerous other works in the literature that equally well support the agreement between

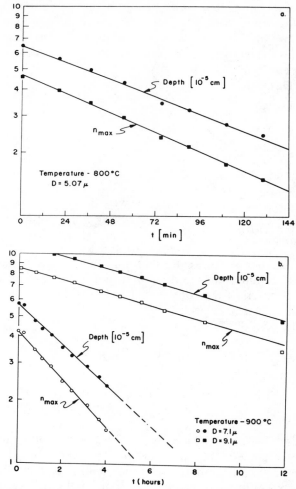

Fig. 7. Semilogarithmic plot of the depth of a sinusoidal profile versus annealing time at two different temperatures; the sample was a Ni(110) single crystal annealed in ultra-high vacuum (Bonzel and Gjostein, 1968a).

phenomenology and experiment. Thus we conclude that mass transfer experiments indeed provide an excellent means of measuring surface (and, quite often simultaneously, bulk) diffusion coefficients (see also Johnson, 1969).

Another important general result is certain empirical correlations for surface diffusion coefficients of different metals. These have been derived on the basis of a comparative study by Gjostein (1967) for fcc and bcc metals. Here the logarithm of D_s was plotted for many metals versus T_m/T, where T_m is the absolute melting temperature of the metal (see Fig. 8). In the case of fcc metals the temperature dependence of D_s is not described by a single activation energy, but the activation energy increases with increasing temperature. This dependence is approximated by two functions (Fig. 8),

$$D_s = 740 \exp(-\varepsilon_1 T_m/RT), \qquad 0.77 < (T/T_m) < 1 \qquad (65)$$

$$D_s = 0.014 \exp(-\varepsilon_2 T_m/RT), \qquad (T/T_m) < 0.77 \qquad (66)$$

where $\varepsilon_1 = 30 \, \text{cal/mole} \, °K$ and $\varepsilon_2 = 13 \, \text{cal/mole} \, °K$.

The temperature dependence of D_s for bcc metals can be correlated in a similar fashion, as seen in Fig. 9. Recent measurements of D_s for W(100) by Bowden and Singer (1969) and for Nb and Mo by Allen (1972) indicate an increase in the apparent activation energy near the melting point, quite analogous to the behavior of fcc metals. Therefore we propose two equations

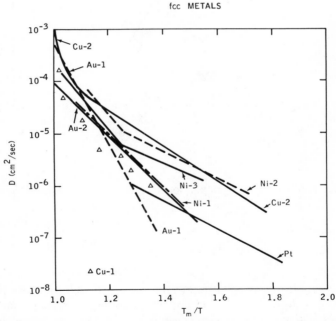

Fig. 8. Arrhenius plot of surface self-diffusion coefficients for fcc metals; T_m is the absolute melting temperature of a metal. Cu-1, Gjostein (1961); Cu-2, Bonzel and Gjostein (1969), Bonzel (1970); Ni-1, Maiya and Blakely (1967); Ni-2, Bonzel and Gjostein (1969); Ni-3, Mills *et al.* (1969); Au-1, Gjostein (1967); Au-2, Henrion (1972); Pt, Blakely and Mykura (1962).

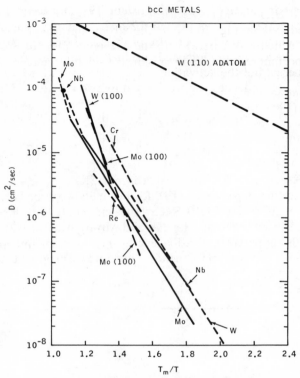

Fig. 9. Arrhenius plot of surface self-diffusion coefficients for bcc metals; T_m is the absolute melting temperature of a metal. Mo, Allen (1972); Mo(100), Singer (1970); Cr, Allen (1969); Nb, Allen (1972); W(100), Bowden and Singer (1969); Re, Allen (1972); W, Barbour *et al.* (1960); W(110) adatom, Ehrlich and Hudda (1966).

to describe the temperature dependence of D, based on the data presented in Fig. 9.

$$D_s = 3.2 \times 10^4 \exp(-\varepsilon_1' T_m/RT), \qquad 0.75 < (T/T_m) < 1 \qquad (67)$$

$$D_s = 1.0 \exp(-\varepsilon_2' T_m/RT), \qquad (T/T_m) < 0.75, \qquad (68)$$

with $\varepsilon_1' = 35$ cal/mole $^\circ$K and $\varepsilon_2' = 18.5$ cal/mole $^\circ$K (Gjostein, 1967). It can also be noted in Fig. 9 that the extrapolated intrinsic diffusivity on a W(110) surface is several orders of magnitude above the mass transfer diffusivity.

The fact that the apparent activation energy of surface diffusion Q_s increases with increasing temperature can also be demonstrated by plotting Q_s/T_m versus $\bar{T}/T_m$, where $\bar{T}$ is a mean temperature representative of the measuring range for the particular value of Q_s. Such a plot was shown by

Neumann and Hirschwald (1972) for a large number of fcc and bcc metals and is reproduced in Fig. 10. Note the strong upturn of the curve for $\bar{T}/T_m > 0.75$, which is consistent with the Arrhenius plots of Figs. 8 and 9.

A third important result in this area is the considerable agreement between the theoretically calculated diffusion coefficients for Cu single crystal surfaces by Wynblatt and Gjostein (1968, 1970) and the experimental data by Bonzel and Gjostein (1969, 1970) for a Cu(110) surface. The details of the theoretical calculations were described in Section III. These data were used to calculate the temperature dependence of D, for example,

$$\text{Adatom:}\quad D_{(110)} = 6.2 \times 10^{-3}\exp(-15{,}200/RT) \qquad (69)$$

$$\text{Vacancy:}\quad D_{(110)} = 2.45 \times 10^{-3}\exp(-18{,}700/RT). \qquad (70)$$

Figure 11 shows the temperature dependence according to Eq. (69) and (70) and also for Cu(111) adatom diffusion together with some experimental data. It can be noted that, at lower temperatures, there is fair agreement between the experimental Cu(110) data and the calculated Cu(110) adatom

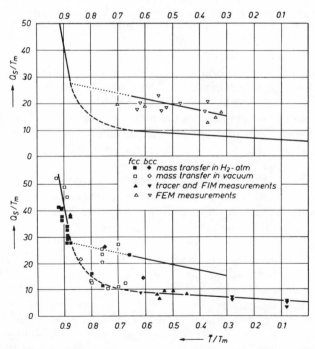

Fig. 10. Plot of experimentally determined activation energies Q_s for surface diffusion versus $\bar{T}/T_m$, where $\bar{T}$ is an average temperature representing the range of validity of Q_s; T_m is the absolute melting temperature of a metal (Neumann and Hirschwald, 1972).

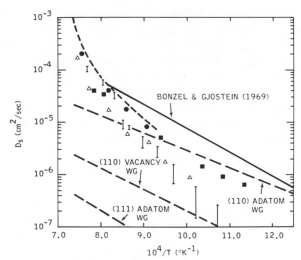

Fig. 11. Arrhenius plot of surface diffusion coefficients for copper. Theoretical data by Wynblatt and Gjostein (1968) are included for comparison. Δ, Gjostein (1961); ●, Collins and Shewmon (1966); ■, Bradshaw *et al*. (1964); I, Bonzel and Gjostein (1968); ---, Bonzel (1970); WG, Wynblatt and Gjostein (1968).

diffusion data. This comparison suggests that vacancy diffusion on a (110) surface may be negligible compared to adatom diffusion.

Furthermore, there is an increasingly larger gap between experimental and theoretical data as one approaches the melting temperature. This behavior was interpreted in terms of dimer and trimer diffusion at high temperature (Bonzel, 1970) (see Section IV). If the line labeled "(110) Adatom" in Fig. 11 characterizes localized adatom diffusion on Cu(110), then there should be contributions due to nonlocalized diffusion of adatoms and higher order defects at high temperature. These contributions were calculated by Bonzel (1970) for a Cu(110) surface and are shown in Fig. 12. The curve labelled D_s describes approximately the measured data of Bonzel and Gjostein (1969) shown in Fig. 11.

In contrast to the large number of mass transfer surface diffusion experiments, there have been only a modest number of papers published dealing with intrinsic surface diffusion. As indicated in Table IV, there are essentially two experimental techniques permitting the measurement of intrinsic diffusivities. The preferred technique is by field ion microscopy, as suggested by Müller (1957) and carried out for the first time by Ehrlich and Hudda (1966) for adatom diffusion on various low-index planes of tungsten.

Ehrlich and Hudda (1966) verified experimentally that the mean square displacement of adatoms was indeed a linear function of the diffusion time, according to Eq. (22). Figure 13 shows an example of this dependence for a W(110) surface at 308 °K. Thus the statistical condition for random walk is satisfied, and a diffusion coefficient can be calculated. Furthermore, the time interval had to be small enough that only a small number of boundary interactions for the diffusing atom could occur. This is necessary for Eq. (22) to hold, since one is dealing with a flat terrace of very limited proportions.

The results of self-diffusion measurements by Ehrlich and Hudda (1966), Bassett and Parsley (1970), and Ayrault and Ehrlich (1972) are listed in Table V. The quantitative agreement between the two sets of data for (110), (321), and (211) W surfaces is remarkably good. It is also noteworthy that diffusion on the (211) plane is faster than on the almost close-packed (110) plane, in contrast to what one would expect on the basis of simple energetic arguments. Theoretical calculations of migration energies by Ehrlich and Kirk (1968) and Wynblatt and Gjostein (1970) have shown that lattice relaxation around the diffusing adatom may substantially influence the calculated diffusion rate. The theoretical values of ΔH_m are included in Table V for comparison with experiment, and it is obvious that agreement is rather

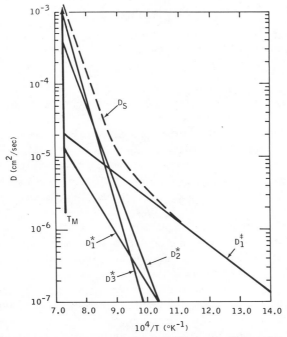

Fig. 12. Arrhenius plot of calculated surface self-diffusion coefficients for copper. The curve labeled D_s describes fairly well the band of experimental data shown in Fig. 11. D_s is the sum of nonlocalized diffusivities $D_i{}^*$ and the localized adatom diffusivity $D_1{}^{\ddagger}$ (Bonzel, 1970).

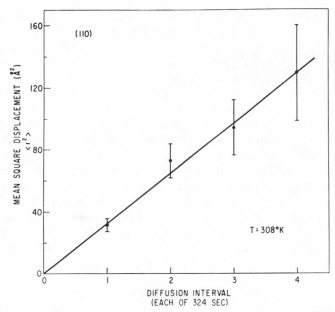

Fig. 13. Plot of the mean square diffusion distance of adatoms on a W(110) surface versus increasing diffusion intervals. The linear dependence is required in order to evaluate the data according to Eq. (22) (Ehrlich and Hudda, 1966).

poor but that the trend in activation energies is reproduced by the data of Wynblatt and Gjostein (1970).

It is quite interesting that adatom diffusion on the close-packed Rh(111) surface seems to proceed with great speed and low activation energy, qualitatively consistent with the calculations for Cu(111) adatom diffusion. It remains to be seen whether diffusion on other low-index Rh surfaces will be slower than on the (111) surface, or whether a similar inversion will be observed as in the case of the bcc metal W.

It would be very desirable to compare for a certain metal intrinsic diffusion coefficients measured by different techniques. The only other technique besides the FIM technique is based on the growth of metallic whiskers. Unfortunately, quantitative whisker growth experiments have been performed only on low melting temperature metals, whereas FIM observations of adatom random walk are available for metals with very high melting temperatures. Even though Ti whiskers have been grown in situ in a field emission microscope by Melmed and Gomer (1961), these have not been utilized for surface diffusion measurements of the type A1 (Table IV) nor for quantitative whisker growth experiments.

TABLE V

EXPERIMENTAL AND THEORETICAL ADATOM DIFFUSIVITIES

Metal	Orientation	D_0(cm^2/sec)	Q(kcal/mole)	Reference
W	(110)	3×10^{-2}	22	Ehrlich and Hudda (1966)
	(321)	1×10^{-3}	20	Ehrlich and Hudda (1966)
	(211)	2×10^{-7}	13	Ehrlich and Hudda (1966)
W	(110)	2.1×10^{-3}	19.9	Bassett and Parsley (1970)
	(321)	1.2×10^{-3}	19.4	Bassett and Parsley (1970)
	(211)	3.8×10^{-7}	13.0	Bassett and Parsley (1970)
Rh	(111)	—	5.5	Ayrault and Ehrlich (1972)
W/Ta[a]	(110)	4.4×10^{-2}	17.9	Bassett and Parsley (1970)
	(321)	1.9×10^{-5}	15.4	Bassett and Parsley (1970)
	(211)	0.9×10^{-7}	11.2	Bassett and Parsley (1970)
W/Re[a]	(110)	1.5×10^{-2}	23.9	Bassett and Parsley (1970)
	(321)	4.8×10^{-4}	20.4	Bassett and Parsley (1970)
	(211)	1.1×10^{-2}	20.3	Bassett and Parsley (1970)
W/Ir[a]	(110)	8.9×10^{-5}	18.0	Bassett and Parsley (1970)
	(211)	2.7×10^{-5}	13.4	Bassett and Parsley (1970)
W/Mo[a]	(211)	2.4×10^{-6}	12.8	Bassett and Parsley (1970)
W/Pt[a]	(110)	$\sim 10^{-4}$	~ 14.0	Bassett and Parsley (1970)
W/Re[a]	(110)	—	20.2	Tsong (1972)
	(321)	—	20.2	Tsong (1972)
W	(110)	—	12.4	Wynblatt and Gjostein (1970)
	(112)	—	9.9	Wynblatt and Gjostien (1970)
	(110)	—	10.1–10.8	Ehrlich and Kirk (1968)
	(112)	—	18.0–25.0	Ehrlich and Kirk (1968)
	(123)	—	20.9–25.0	Ehrlich and Kirk (1968)

[a] Second element is diffusing adatom.

Good quantitative data on whisker growth were recently published by Dittmar and Mennicke (1970, 1972) for K and Rb. Here one should remember that the primary quantity measured is the mean root square diffusion distance of adatoms (or whatever equivalent might diffuse). This quantity $\bar{x}$ is plotted as a function of reciprocal temperature in Fig. 14 for K and Rb. Because of Eq. (61), $\bar{x}$ is related to the intrinsic surface diffusion coefficient or

$$\bar{x} = \text{const} \times \exp\left(\frac{\Delta H_{\text{des}} - \Delta H_{\text{m}}}{2RT}\right), \tag{71}$$

where ΔH_{des} is the binding energy of the diffusing adatom (or defect). The

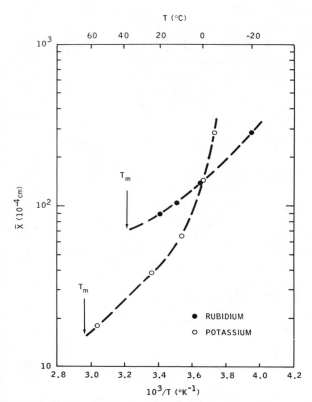

Fig. 14. Arrhenius plot of the mean square diffusion distance of adatoms on whisker surfaces. The strong curvature, particularly in the case of potassium, suggests a change in diffusion mechanism (Dittmar and Mennicke, 1970, 1972). Note that here $\bar{x}$ is only a function of temperature and not time, contrary to the data in Fig. 13.

interesting feature in Fig. 14 is the noticeable curvature in the dependence of $\bar{x}$ on $1/T$, particularly in the case of K. Since ΔH_{des} is generally not expected to vary in the range of adatom coverages of interest, the variation in the activation energy must result from a variation of ΔH_m. It is conceivable that, because of the supersaturation of adatoms in these experiments, the concentration of dimers and trimers may also be high and that these may contribute to the overall diffusion rate, especially at temperatures near the melting point where nonlocalized diffusion can take place.

For Rb the increase in ΔH_m from low temperature to high temperature is approximately 2.86 to 6.18 kcal/mole based on the data in Fig. 14 and on $\Delta H_{des} \simeq 0.6\,\Delta H_s = 12.3$ kcal/mole. For K one obtains $\Delta H_m = 7.0$ kcal/mole at high temperature, $\Delta H_{des} = 13.1$ kcal/mole; but at low temperature the difference $\Delta H_{des} - \Delta H_m$ is larger than the estimated ΔH_{des}, a fact which is not well understood at this time.

Mennicke and Dittmar (1972) evaluate surface diffusion coefficients of the order of 1 to $3\,cm^2/sec$ that are larger than gaseous diffusion coefficients. Hruska and Hirth (1961) have discussed the reliability of D values determined from whisker growth experiments and have pointed out that smaller diffusivities should more likely be observed. In this context it is also important to be aware of impurity effects which, for example, could sufficiently alter ΔH_{des} and thus D, because the product $D\omega$ is measured.

B. Surface Self-Diffusion on Metals in the Presence of Adsorbates

In the foregoing section we have dealt with surface self-diffusion measurements on *clean* metal surfaces, where the meaning of "clean" was directly related to the analytical technique of assessing surface cleanliness. It is well known, however, that adsorbed impurities may cause a suppression or an enhancement of surface diffusion depending on the kind of impurity. These effects are illustrated in Fig. 15 in the case of Cu surface diffusion. As the

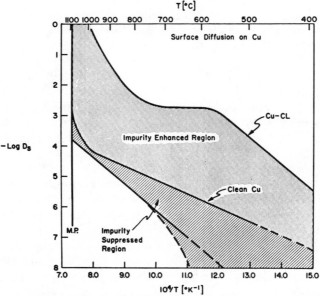

Fig. 15. Arrhenius plot of surface self-diffusion coefficients for copper summarizing the possible impurity effects on D_s. The upper curve represents the largest D_s values observed so far for Cu in the presense of Cu_2Cl_2 vapor (Delamare and Rhead, 1971). The curve for clean Cu was taken from the data by Bonzel and Gjostein (1969) and Bonzel (1970). The lower curve is an anticipated boundary based on a large variety of measurements (Bonzel, 1973).

data representing a clean surface, we have taken those of Bonzel and Gjostein (1969) where the impurity suppressed region is due to C, Ca, Mn, O_2, and possibly others (Bonzel, 1973); the impurity enhanced region is here caused by Cl and taken from measurements by Delamare and Rhead (1971).

In accordance with the experimental findings there are then two classes of impurities:

(a) Those having a higher melting point (cohesive energy) than the substrate; they will most likely cause a decrease in the rate of surface self-diffusion.

(b) Those having a lower melting point than the substrate; their presence on the surface can cause a substantial increase in D_s where the magnitude of this increase generally depends on the coverage of the impurity.

Experiments by Bonzel and Gjostein (1969), Olsen et al. (1972), and Pichaud and Drechsler (1972) have demonstrated that carbon is a very effective impurity of class (a) for the metals Cu, Au, and W. Since carbon is a likely contaminant in all oil diffusion pump systems, it is understandable that many early measurements of surface self-diffusion of metals suffered from unclean environments and hence showed poor reproducibility. Bonzel and Gjostein (1967) showed in a quantitative model how adsorbed impurities can poison kink and ledge sites, reduce the concentration of diffusing defects, and thus suppress the rate of surface self-diffusion.

On the other hand, Perdereau and Rhead (1967), Henrion and Rhead (1970), Delamare and Rhead (1971), and Pichaud and Drechsler (1973) have demonstrated convincingly that a large group of elements of type (b) exist that lead to very sizable increases in surface diffusion rates. The most striking example is perhaps the Cu–Cl system (Fig. 15), where surface self-diffusion of copper in the presence of adsorbed chlorine is enhanced by a factor of $\sim 10^4$ (Delamare and Rhead, 1971). The largest diffusivities recorded are of the order of $1 \, cm^2/sec$ and thus comparable to those reported by Mennicke and Dittmar (1972) in connection with Rb whiskers. In the case of the W–Ni system, a decrease in activation energy from 71 kcal/mole to 21 kcal/mole is found if the Ni coverage is increased to about one monolayer (Pichaud and Drechsler, 1973).

This well-documented surface diffusion enhancement effect is difficult to explain from a mechanistic point of view. So far two different ideas have been advanced. The first by Rhead (1969) is based on a "surface melting" phenomenon that is supposed to be equivalent to general disorder at the surface resulting in a high concentration of diffusable defects. The second model was proposed by Bonzel and Gjostein (1970) and is based on the concept of nonlocalized diffusion of defects. In the presence of an adsorbate A, it is possible that complexes of the kind $AB, A_2 B_2$, or generally $A_x B_y$

(with B as the substrate species) form corresponding to dimers, trimers, etc. in the clean metal case. Because of the high concentration of A, there would be a high probability for forming these complexes, and hence a high diffusivity may be expected. Since the species A accelerates the diffusion and stays on the *surface* of B, the resulting diffusivity was termed a "catalytic diffusivity" by Bonzel (1973).

In the case of the Cu–Cl system, however, it is also conceivable that, because of the high vapor pressures of copper halides, transport through the vapor phase might not be negligible, contrary to the argument given by Delamare and Rhead (1971). Since there is a dynamic equilibrium between copper and $Cu_x Cl_x$ vapor without resulting in a net deposition (or net loss) of Cu, the conditions are favorable for transport of Cu in the form of $Cu_x Cl_x$. For vapor pressures of the order of 10^{-2} to 10^{-1} Torr, very large rate constants for this process [c.f. Eq. (45)] are calculated indeed. This, however, is not consistent with the finding of an average slope of 0.26 ± 0.3 for the kinetics of grooving (Delamare and Rhead, 1971). It seems that more experimental evidence is needed at this point in order to decide for or against vapor transport as an active mode for grain boundary grooving in the presence of copper halides.

C. Heterogeneous Surface Diffusion

Many measurements of heterogeneous surface diffusion have been carried out by FEM and radioactive tracer techniques (B3 and D1 of Table IV). For surface diffusion of gaseous molecules (or atoms) on FEM tips, there are according to Gomer (1959) three types of diffusion observed.

(a) For initial deposits in excess of a monolayer, diffusion occurs at very low temperature with a sharp boundary. This behavior is indicative of diffusion in second and higher layers of physically adsorbed gas on top of the chemisorbed layer ("unrolling carpet"). The corresponding activation energies are very low.

(b) For initial coverages of 0.3–1.0 monolayers, diffusion occurs at medium to high temperatures with a sharp boundary. This type of diffusion is characteristic of surface diffusion of the chemisorbed species for relatively high concentration, and therefore the measured activation energies are high.

(c) For smaller coverage than that necessary for type (b) diffusion or after its cessation, diffusion is observed without a sharp boundary at higher temperature. The corresponding activation energy is slightly higher than for the type (b) process. Here again one is dealing with surface diffusion of the chemisorbed species but for lower coverages.

320   H. P. BONZEL

Table VI lists data for all three processes in the cases CO and O_2 on W
and Pt tips. In addition results are shown for diffusion of H_2 on W and Pt
in the chemisorbed layer. The same technique has been used to study the
diffusion of a metal A on substrate B. Examples of such measurements are also
shown in Table VI. Many of these measurements demonstrate very clearly
the effect of anisotropy and of coverage dependence in surface diffusion.

TABLE VI

HETEROGENEOUS SURFACE DIFFUSION MEASUREMENTS BY FEM TECHNIQUES

Substrate	Diffusing species	Orientation of substrate	Q(kcal/mole)	References
W	CO	(110)	[60.0]	Gomer (1959)
W	CO	(110)[b]	36.0	Gomer (1959)
W	O	—	30.0	Gomer (1959)
W	O	(110)[b]	24.8	Gomer (1959)
W	O	(100)[b]	22.7	Gomer (1959)
W	H	—	9.6–16	Gomer (1959)
W	H	(110)[b]	5.9	Gomer (1959)
CO_2/W[a]	CO_2	—	2.4	Gomer (1959)
CO/W[a]	CO	—	[0.9]	Gomer (1959)
O/W[a]	O_2	—	0.9	Gomer (1959)
W	Xe	—	[3.0]	Gomer (1959)
W	Kr	—	[0.9]	Gomer (1959)
W	Ar	—	0.6	Gomer (1959)
Ni	H	—	7.0	Gomer (1959)
Pt	CO	—	14–15	Lewis and Gomer (1967)
Pt	CO	(111)[b]	10.1	Lewis and Gomer (1967)
Pt	O	—	34.0	Lewis and Gomer (1968)
Pt	O	(111)[b]	27.0	Lewis and Gomer (1968)
Pt	H	—[b]	4.5	Lewis and Gomer (1969)
CO/Pt[a]	CO	—	0.9	Lewis and Gomer (1967)
O/Pt[a]	O_2	—	1.4	Lewis and Gomer (1968)
W	Cu	(100)	22.0	Melmed (1965)
W	Cu	(110)	<8.0	Melmed (1965)
Cu/W[a]	Cu	—	9–13	Melmed (1965)
W	Au	(101)	14–34	Vladimirov and Sokolskaya (1970)
W	Ti	(011)	23–38	Vladimirov et al. (1970a)
W	Yb	(011)	10.4	Marinova et al. (1970)
W	Nd	(101)	17.2	Marinova et al. (1970)
W	Y	(113)	44–55	Palynkh et al. (1971)

[a] A/B means: B covered with A.
[b] Boundary diffusion.

It is somewhat surprising to see that the activation energies of hetero-geneous surface diffusion are quite low compared to the energy of surface self-diffusion of the substrate material. This is particularly evident in the case of W, which has been used most frequently as a substrate. These findings are not unique to FEM measurements, since the same behavior was found in the analysis of results obtained by the tracer technique (on flat macro-scopic samples). The latter results are summarized in Table VII, together with some measurements by other techniques.

Abramenkov et al. (1971) recently measured surface diffusion of Cu on a Mo ribbon employing secondary ion emission as an analytical tool for measuring the concentration curves after diffusion. Their results, although obtained in a relatively poor vacuum, are very similar to earlier FEM results by Melmed (1965) for diffusion of Cu on W.

Roulet (1973; Roulet and Borel, 1972) investigated the diffusion of Ag on low-index as well as vicinal Cu single crystal surfaces. In these experiments Ag was continuously deposited through a mask onto the Cu surface, while at the end of a run the concentration profile of Ag was monitored by interference microscopy after the surface had been oxidized in air at 250° C. The scatter in diffusion coefficients determined in this fashion is quite ap-preciable, as seen in Figs. 16 and 17. It is not clear whether this is due to a flux dependence, slight variations in the oxidation technique, differences in surface preparation, or other causes. Roulet (1973) unfortunately reports average D_0 factors and activation energies (see Table VII), but the actual data in Fig. 16, for example, suggest strongly that the activation energy for Ag diffusion on Cu(110) is only about ≈ 5 kcal/mole for $T \geqslant 300°$ C and that below 300° C an apparent suppression effect sets in. A similar behavior may be present on the Cu(111) surface (Fig. 17), where for $T \geqslant 400°$ C almost no temperature dependence of the surface diffusivity is observed. This in-terpretation would be much more consistent with the theoretical work of Wynblatt and Gjostein (1968).

It is also interesting that the activation energies for vicinal Cu surfaces as measured by Roulet (1973) are lower than those for low-index surfaces but that the average diffusivities are of the same order of magnitude. Even though the technique and the results obtained are quite interesting, the large scatter in the data (factor of 10 quite common!) does not yet permit one to draw solid conclusions. Therefore it does not appear to be justified to forward a rationale of the average diffusivities and average activation energies in terms of ledge loops ("Lochkeime") on the surface, as proposed by Roulet (1973).

Most of the tracer surface diffusion data recently reviewed by Suzuoka (1965) and Gjostein (1970a) show a low activation energy compared to that obtained by mass transfer diffusion. Table VII summarizes the available data, and with the exception of the measurements by Choi and Shewmon

TABLE VII

SURFACE DIFFUSION DATA OBTAINED BY RADIOACTIVE TRACER AND RELATED TECHNIQUES

Adsorbate/substrate	Technique	$D_0(\mathrm{cm^2/sec})$	Q (kcal/mole)	Reference
Ag/Ag	Tracer	0.2	10.3	Nickerson and Parker (1950)
Ag/Ag	Tracer	0.13	12.0	Ikeda (1965)
Ag/Ag	Tracer	—	—	Winegard and Chalmers (1952)
Ag/Ag	Tracer	—	19.9	Gal et al. (1970)
Ag/Ag	Tracer	0.3	11.8	Geguzin and Kovalev (1963)
Ag/Ni	Tracer	22.4	16.0	Geguzin and Kovalev (1963)
Ag/Ge	Tracer	$(0.9–5.4) \times 10^{-4}$	8.5	Kosenko and Khomenko (1961)
Ag/Ag	Tracer	1.5×10^{-5}	8.1	Drew and Pye (1963)
Au/Au	Tracer	—	10–40	Geguzin et al. (1964)
Ni/Ni	Tracer	—	13.8	Pye and Drew (1964)
Ni/Ni	Tracer	300	38.0	Wolfe and Weart (1969)
Au/Cu	Tracer	2.7×10^5	51.5	Choi and Shewmon (1964)
Au/Cu	Tracer	—	20–30	Austin et al. (1966)
Ge/Cu	Tracer	5×10^5	45.0	Shewmon and Choi (1964)
W/W	Tracer	1.1×10^{-3}	31.6	Neumann et al. (1966)
Ni/Ni	Tracer (H_2)	10^a	13.9	Gal et al. (1973)
Ni/Ni	Tracer (Vac)	10^{6a}	49.4	Gal et al. (1973)
Cu/Mo	SIE	8.7×10^{-4}	12.5	Abramenkov et al. (1971)
Ag/Cu(100)	VD–O	0.062	17.5	Roulet[b] (1973)
Ag/Cu(111)	VD–O	0.46	17.5	Roulet[b] (1973)
Ag/Cu(110 ⟨110⟩	VD–O	2.8	17.0	Roulet[b](1973)
Ag/Cu(110)(001)	VD–O	7.3	19.6	Roulet[b] (1973)
Ag/Cu(001)[c] ⟨110⟩	VD–O	2.4×10^{-4}	9.7	Roulet[b] (1973)
Ag/Cu(001)[c] ⟨110⟩	VD–O	5.8×10^{-4}	10.3	Roulet[b] (1973)
Ag/Cu(111)[c] ⟨112⟩	VD–O	0.023	14.3	Roulet[b] (1973)
Ag/Cu(111)[c] ⟨110⟩	VD–O	0.016	12.9	Roulet[b] (1973)
Cs/W		0.2	14.0	Langmuir and Taylor (1932)
Ag/Ni	PC	4×10^5	60.0	Kaganovskiy and Ratinov (1971)
Ag/Ni	PC	18.0	37.0	Kaganovskiy and Ratinov (1971)
Ni/W	PC	—	30.0	Geguzin et al. (1972)

SIE = Secondary ion emission.
VD–O = Vapor deposition and subsequent oxidation.
PC = Particle coalescence

[a] D_0 is calculated assuming $a = 2 \times 10^{-8}$ cm for thickness of diffusion layer.
[b] Average values for D_0 and Q.
[c] Vicinal surfaces.

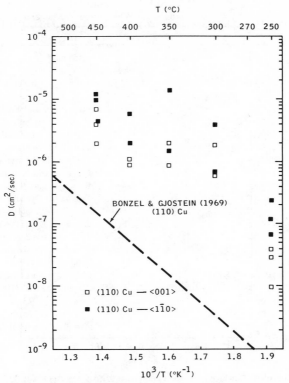

Fig. 16. Arrhenius plot of surface diffusion coefficients for Ag on Cu(110) surfaces (Roulet, 1973). Note large scatter in data but also an apparent suppression effect at 250° C. The average activation energy for $T \geqslant 300°$ C seems to be only 5 kcal/mole.

(1964), Wolfe and Weart (1969), and Gal *et al.* (1973), all apparent activation energies are below 20 kcal/mole. Suzuoka has pointed out the danger of mis-analysis of concentration curves that arises from simultaneous bulk diffusion, but it is not clear yet whether this fact can be responsible for low Q values. Shewmon (1963) and Suzoka (1965) also point out that preferential adsorption of the diffusing material to the surface (segregation) could alter the penetration curves and possibly the temperature dependence of D_s (see also Bonzel and Aaron, 1971; Blakely and Shelton, Chapter 4 of Volume I; Williams and Boudart, 1973).

This raises the general question of whether the activation energies for heterodiffusion should always be lower than for self-diffusion, or whether one measures by these techniques an intrinsic diffusivity rather than a mass transfer diffusivity (see Section V.B). It is possible that in the majority of

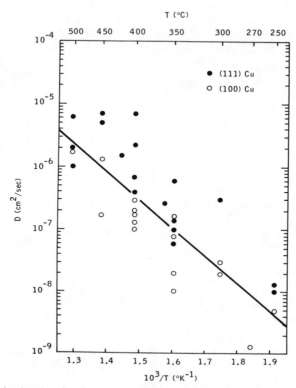

Fig. 17. Arrhenius plot of surface diffusion coefficients for Ag on Cu(111) and Cu(110) surfaces (Roulet, 1973). The data for Cu(111) in the region $T > 350°$ C are systematically higher than those for Cu(100).

these experiments the adatom concentration is not in equilibrium with the substrate and that therefore the measured diffusivity is closer to an intrinsic diffusivity.

In contrast to the previous experiments, where it is not known whether intrinsic or mass transfer diffusion has been studied, Bassett and Parsley (1970) and Tsong (1972) have investigated heterogeneous surface diffusion by the FIM technique. They have looked at the diffusion of Ta, Mo, Re, Ir, and Pt on various single crystal faces of W and found that ΔH_m was quite similar to that of W self-diffusion (at least of the same order of magnitude). This again seems to suggest that the diffusivities measured by FEM, radio-active tracers, etc. yield activation energies representative of intrinsic surface diffusion.

Finally there is one example of particle coalescence (metal on metal) in the literature, namely for the system Ag on Ni measured by Kaganovskiy and Ratinov (1971). The average size of Ag particles in these experiments was large, of the order of 4×10^{-4} cm. The particles were obtained by vapor deposition of a 2000 Å thick Ag film, which was then annealed at 800° C for 0.5 hour. During this anneal the film broke up into particles. The coarsening of these particles was then studied at temperatures of 700° C to 900° C. The average particle radius $\bar{R}$ was measured as a function of time, and according to the formula

$$\bar{R}^4 = \frac{2\gamma a^4}{kT} D_s t, \tag{72}$$

where a is the lattice parameter, the heterodiffusion coefficient D_s was determined. The results for two different Ni grains showed activation energies of 37 and 60 kcal/mole, which are higher than those measured by tracer techniques (see Table VII).

D. Surface Diffusion on Multicomponent Crystals

In the preceding sections we discussed surface diffusion on a one-component substrate where the diffusing species was either of the same type as the substrate (self-diffusion) or a different type (heterodiffusion). If the substrate itself is heterogeneous, i.e., a multicomponent crystal, the process of surface diffusion can be very complex. For example, if the substrate is an AB_2 crystal, we could investigate the diffusion of the species A, B, AB_2, C, $AC, BC, \ldots$ on the surface of AB_2. Other variations of this scheme may be thought of, but this example illustrates the possible complexity of surface diffusion. As a consequence it would be difficult to identify the diffusing defect. Only a relatively small number of experiments in this area have been published; in the following we will briefly report some of the results according to materials investigated.

1. ALLOYS

If a mass transfer technique for measuring surface diffusion is applied to a multicomponent crystal, an effective surface diffusion coefficient will be measured. Blakely and Li (1966) and Li et al. (1966) have extended Mullins' theory to include ionic crystals. A crystal of the type $A_\alpha B_\beta$ with Schottky disorder was considered. With the individual diffusivities D_A and D_B for species A and B, respectively, the effective diffusivity can be written as

$$D_s^{\text{eff}} = \frac{D_A D_B}{\alpha D_B + \beta D_A}. \tag{73}$$

This means that the species with the lower diffusivity will be rate controlling.

Equation (73) was also applied to binary alloys by McLean and Hirth (1969) in their investigation of the Au–Ag system. They particularly checked the validity of Eq. (73) at $T = 950°$ C for two alloys with 54.4 at% and 27.2 at% Ag and found the parabolic dependence predicted by Eq. (73). This agreement, however, was itself dependent on the choice of self-diffusion parameters for pure Au and Ag. Had McLean and Hirth chosen the Au self-diffusion data of Gjostein (1965) rather than their own (McLean and Hirth, 1968), Eq. (73) would not have reproduced the variation of D_s^{eff} with increasing content of Ag.

Mills and Douglas (1970) reported surface diffusion coefficients for Fe–3% Si alloy single crystals of (110) and (100) orientation. The temperature dependence of the effective diffusion coefficients measured between 1150 and 1400° C was described by $\bar{D}_0 \simeq 6 \times 10^4 \, cm^2/sec$ and $\bar{Q} = 58,500 \, kcal/mole$ (average values for both orientations). This diffusivity is lower than the extrapolated self-diffusivity of α-Fe reported by Zahn (1964).

Other surface diffusion studies of alloys have been carried out in thin film couples by Pines et al. (1967, 1969, 1970). However, it is quite difficult to compare the results of these studies with diffusion measurements in bulk crystals because of several important differences, e.g., grain size, defect structure, purity, definition of "surface layer," methods of evaluation. It is quite common that low activation energies are observed by the thin film technique, whereas the D_0 factors are unknown.

2. OXIDES AND CARBIDES

Mass transfer techniques, such as grain boundary grooving, single scratch, and sinusoidal profile decay, have also been used to measure surface diffusion coefficients of oxides and uranium carbide. Table VIII lists the results of these investigations in terms of preexponential factors and activation energies. Note the large preexponential factors $(10^5 - 10^8) \, cm^2/sec$ and activation energies, with the exception of the data on Morganite by Robertson and Chang (1966) and uranium carbide by Nicholas and Hodkin (1971). The data on Morganite have been rationalized in terms of impurity influence on D_s. Morganite contains small amounts of silica and calcia that may segregate at the free surface to form a thin layer with a lower melting point increasing the rate of surface diffusion. This interpretation is consistent with an independent observation that an addition of silica and calcia to the surfaces of the Linde material caused a similar enhancement of D_s (see discussion by Robertson, 1969). The Linde material most likely represents pure alumina with $Q = 130 \, kcal/mole$ for surface diffusion (Robertson, 1969).

The measurements of surface diffusion on UC are slightly controversial.

TABLE VIII

SURFACE DIFFUSION ON MULTICOMPONENT SYSTEMS

Material	Temperature range °C	D_0 (cm²/sec)	Q (kcal/mole)	Reference
Al_2O_3 (Morganite)	1400–1750	9×10^2	75	Robertson and Chang (1966)
Al_2O_3 (Linde)	1400–1700	1×10^8	128	Robertson (1969)
Al_2O_3 (Lucalox)	1100–1750	8×10^8	133	Robertson (1969)
Al_2O_3	1540–1800	5×10^5	110	Shackelford and Scott (1968)
MgO	1200–1500	2.3×10^5	90	Robertson (1967)
MgO	1100–1500	8×10^4	88.5	Henney and Jones (1968)
$UO_{2.005}$	1200–1700	1.3×10^8	110	Henney and Jones (1968)
$UO_{2.00}$	1200–1700	1×10^7	110	Henney and Jones (1968)
$UO_{2.00}$	1550–1700	2.1×10^7	125	Amato et al. (1966)
$UO_{2.00}$	1450–1700	5.6×10^6	119	Reynolds (1967)
$UO_{2.00}$	1544–1915	5.64×10^7	120.7	Gulden (1967)
UC	1100–1400	5.6×10^{-8}	31.4	Nicholas and Hodkin (1971)
Au-54 at% Ag	800–1000	2×10^6	60	McLean and Hirth (1969)
Fe-3 at% Si	1150–1380	6×10^6	58	Mills and Douglas (1970)

The results of Nicholas and Hodkin (1971) were obtained at temperatures of 1100 to 1400° C, while measurements in the range 1800 to 2160° C by Maiya and Routbort (1972) indicated volume diffusion control. The latter was established by measuring the wavelength dependence of the relaxation constants.

It is difficult to understand the process of surface self-diffusion on materials such as Al_2O_3, MgO, UO_2, or UC. Because of the high bond energies of the individual molecules, a dissociation on the surface seems unlikely, and hence a contribution of individual surface diffusion of Al and O ions, for example, may be negligible. But since surface diffusion of molecules such as those mentioned above is taking place at high temperatures, there is a probability for nonlocalized diffusion, and hence high D_0 factors may be expected.

3. ALKALI HALIDES

The TLK model of a crystalline surface was originally proposed by Stranski (1928) for a NaCl crystal, and Volmer (1939) later discussed the possibility of surface diffusion on NaCl, pointing out that the energy of migration should be very low, of the order of 4 kcal/mole. Early qualitative experiments on surface diffusion of NaCl were carried out by Gyulai (1935) and Neumann (1938). However, it was only recently that more quantitative experiments were undertaken, mostly in the context of vapor deposition or by using radioactive tracers. Table IX is a summary of the available data.

TABLE IX

SURFACE DIFFUSION ON ALKALI HALIDES

Substrate	Diffusing species	Technique	D_0 (cm^2/sec)	Q (kcal/mole)	Reference
NaCl	Pt	VD	—	4.2	Sumner (1965)
NaCl	^{22}Na	Tracer	—	11.0	Geguzin and Kaganovskii (1969)
NaCl	^{36}Cl	Tracer	—	8.0	Geguzin and Kaganovskii (1969)
NaCl	^{110}Ag	Tracer	—	16.0	Geguzin and Kaganovskii (1969)
KBr	^{82}Br	Tracer	—	9.5	Geguzin and Kaganovskii (1969)
KCl	Au	VD	—	2.8	Trofinov et al. (1969)
NaCl	Au	PC	10^{-11}	14.0	Geguzin et al. (1969)
NaCl	Au	VD	—	3.5–10.5[a]	Hayek and Schwabe (1972)
AgBr	^{110}Ag	Tracer	2×10^{-4}	7.8	Tan et al. (1971)
		IC	—	8.1	Tan et al. (1971)

VD = vapor deposition; PC = particle coarsening; IC = ionic conductivity.
[a] $T > 300°$ C.

At a glance we recognize that the measured activation energies for surface diffusion are indeed very low, ranging from about 3 to 16 kcal/mole. There does not seem to be a large difference between self-diffusion and hetero-diffusion, and, furthermore, the diffusion of cations and anions is described by nearly the same activation energy. This latter observation may imply that the diffusion of individual ions probably does not occur but that rather more neutral molecules diffuse at the surface. Volmer (1939) suggested an alternate hopping of cation and anion, as illustrated in Fig. 18.

In a qualitative sense fast surface migration of adsorbed species on alkali halides is also supported by the fact that decoration of ledges by Au, Pd, etc. on these crystals is quite simple to achieve (Bassett, 1958; Bethge, 1964). Epitaxial growth of Au, Ag, and other metals on alkali halides or similar substrates occurs quite readily at low temperatures, as reported by Pashley et al. (1964) and Rhodin et al. (1969). Surface diffusion of the adsorbed species plays an important role during the initial deposition and nucleation phase but also during later secondary growth stages (particle coarsening, Ostwald ripening). However, there is still a considerable lack of quantitative experiments permitting an extraction of surface diffusion coefficients. Furthermore, there exists a controversy in this field as to whether small

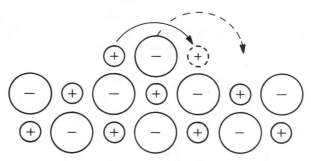

Fig. 18. Schematic illustrating a surface diffusion mechanism on a NaCl surface (Volmer, 1939). Anion and cation hop alternately, thus giving rise to molecular diffusion.

particles grow by an Ostwald ripening mechanism or by self-migration of particles, collision, and coalescence. Some evidence for "Brownian motion" of small particles is discussed in the following section.

E. Brownian Motion of Particles on Substrate Surfaces

Bassett (1961, 1964) first reported on the motion of individual Cu and Ag particles on substrate surfaces such as graphite and molybdenum sulfide. This motion apparently consisted of translations as well as rotations of particles. More recently Masson and Kern (1968) and Masson *et al.* (1970, 1971, 1972) have experimentally verified the migration of individual particles ranging in size from ∼ 25–100 Å.

Masson and Kern (1968) originally studied the coalescence of Au particles on a Si(111) surface and found that the growing (large) particles having an average size of about 0.1 μ were surrounded by small particles of 20 to 60 Å diameter, but only in their immediate vicinity. This observation is in conflict with an Ostwald ripening process, where one would expect that large particles should be surrounded by an area of very low particle concentration. On the other hand, Masson and Kern argued that under certain assumptions a particle distribution such as the one observed may be obtained if the small particles are allowed to undergo Brownian motion. It appears that an attractive force causes the small particles to cluster around the large particle but that actual coalescence does not occur to any appreciable extent (interface limited). This fact is supported by the size distribution of particles, which shows very little change before and after diffusion.

In a series of experiments Masson *et al.* (1971, 1972) showed that migration of particles occurs, but coalescence is almost negligible. Because of the wealth of evidence in favor of individual particle migration, a particle diffusion coefficient is evaluated. For the diffusion of Au particles on a KCl(100) substrate, one finds in the temperature range 60–130° C for D_0 factors and activation energies Q:

$$d \approx 31 \text{ Å} \quad D_0 = 10^2\text{--}10^6 \text{ cm}^2/\text{sec}$$
$$Q = 26\text{--}30 \text{ kcal/mole}$$
$$d = 24 \text{ Å} \quad D_0 = 10^{-4}\text{--}1 \text{ cm}^2/\text{sec}$$
$$Q = 14\text{--}18 \text{ kcal/mole},$$

where d is the average diameter of the particles.

Masson *et al.* point out that particle migration may be due to self-diffusion on the particle surface or to Brownian motion of the particles. On the basis of some quantitative arguments the authors decide in favor of Brownian motion. It is also observed that the migration distance of particles has a maximum that is reached after some time. Furthermore, this maximum in migration distance indicates the moment where the migrating particle finds an epitaxial position on the surface.

Masson *et al.* propose as a consequence of these experimental results on particle migration that substrate temperature is a very important parameter for epitaxial growth. They argue that in order to obtain good epitaxy the temperature of the substrate should be high enough for Brownian motion of particles to occur, so they can translate and rotate into an epitaxial position before they collide and coalesce.

F. Surface Diffusion of Adsorbed Molecules on Porous Substrates

In most of the surface diffusion measurements discussed so far the surface itself was fairly well defined and characterizable in terms of structure, composition, and sometimes cleanliness. However, there is also a class of substrates on which surface diffusion has been measured but whose surfaces are not well characterized. These are porous materials, such as SiO_2, Al_2O_3, and zeolites. For example, the diffusion of Ar on porous SiO_2 surfaces was studied by Boddenberg (1970) at 78 and 90° K by an equilibrium technique. Typical diffusion coefficients are of the order of $10^{-5} \text{ cm}^2/\text{sec}$. Boddenberg *et al.* (1970) also measured surface diffusion of benzene and cyclohexane on porous SiO_2 by NMR techniques and found $D \simeq 10^{-5} \text{ cm}^2/\text{sec}$ at 30° C. These values vary considerably with coverage and usually show a maximum near monolayer coverage.

We will not discuss these results in detail but rather refer to review articles by Carman (1956), Barrer (1967), and Haul and Boddenberg (1969).

VIII. CONCLUSION

In this review of surface diffusion we have pointed out many different areas in which more or less extensive experimental research has been carried out. In most cases the corresponding phenomenology is well understood and only minor approximations are necessary in order to find useful mathematical equations that lend themselves to an evaluation of the experiment. On the other hand, the theory of surface energetics is still in an underdeveloped state, mainly for two reasons:

(1) Generally a pairwise interaction model is used to calculate formation and migration energies and entropies of surface defects, a questionable approach when applied to solid surfaces.

(2) Even though this model may have limited validity, not enough calculations[†] have been performed in order to assess the degree of agreement with experiment.

Consequently, one would wish to see more activity in the field of experimental investigation of surface diffusion on clean and well characterized surfaces as well as in the field of theory, both in the framework of the pairwise interaction model and quantum theory, such that a more extensive comparison of experiment and theory could be made.

In close context with the foregoing it should be said that the treatment of high temperature surface diffusion by statistical thermodynamics methods is still fairly crude, especially since again reference is made to a surface potential calculated by the nearest neighbor model. Experimentally, the effect of "curvature" in the Arrhenius plot of fcc and bcc metals seems well established. However, there may be one effect that has not yet been properly taken into account and that could cause an enhancement of, e.g., the decay of a sinusoidal profile. This is the process of free evaporation that takes place in a vacuum system, in contrast to vapor transport under thermodynamic equilibrium conditions. Attempts at accounting quantitatively for this effect are presently being undertaken by Drechsler (1973), but it is doubtful that this effect will be significant. It was shown by Winterbottom (1969), in the case of Ag, that the rate of vaporization is independent of the undersaturation in the vapor phase. Therefore free evaporation and evaporation under thermodynamic equilibrium are indistinguishable for Ag. However, the latter effect has usually been estimated during the evaluation of high temperature surface diffusion experiments and amounts in the case of Cu to no more than 0.5% above 1300° K for a periodicity of 10 μ.

[†] Including lattice relaxations (Wynblatt and Gjostein, 1970).

The high temperature effect of surface diffusion of metals, represented by an increasing activation energy with increasing temperature and by diffusion coefficients as high as 10^{-3} cm^2/sec, indicates that maximum surface diffusion rates are higher than those for grain boundaries, dislocations, and even liquids (Gjostein, 1973). This means that indeed the third dimension at the solid surface must play an important role. This thought is amplified by the data for surface diffusion in the presence of adsorbates, where diffusivities of the order of 1 cm^2/sec are observed. These are in the range of gaseous diffusion constants, and therefore it is fair to say that the surface diffusion process must be similar to that in the gaseous phase (Bonzel and Gjostein, 1970). This fact is further supported by the whisker growth experiments cited in Section VII.A, where surface diffusivities of 1 cm^2/sec were reported. Here one has a supersaturated adatom concentration and hence favorable conditions for pair and cluster formation. It should be borne in mind, however, that a complete separation of reevaporation and surface diffusion of adatoms has not been achieved as yet. The information to be gained from such a separation would be helpful for a better understanding of surface diffusion near the melting temperature of a metal.

It becomes quite apparent in this chapter that not too many experimental data are available in the areas of particle coarsening, Brownian motion of particles, surface diffusion during vapor deposition, and surface diffusion on multicomponent materials. This is most likely due not only to experimental difficulties but also indicative of some lack of theoretical understanding of these processes. One certainly would like to see more activity toward these interesting directions. At the same time it would be desirable to perform additional experiments on well characterized clean metal surfaces, in particular with respect to the question of anisotropy of surface diffusion. Not a single clean mass transfer experiment in this area is available, and yet the necessary techniques for performing anisotropy experiments on clean single crystal surfaces are well known and proven.

A continued study of the process of surface diffusion on solid surfaces is of great importance toward a fundamental understanding of dynamic surface processes. Together with investigations of surface structure, adsorption, and catalysis, the study of surface diffusion will provide information on a centrally important function, the surface potential. One should hope, therefore, that research efforts in this area will be intensfied in the near future, particularly along the general directions outlined above.

REFERENCES

ABRAMENKOV, A. D., SLEZOV, V. V., TANATAROV, L. V., and FOGEL, YA. M (1971). *Fiz. Tverd. Tela* **12**, 2934.
ALLEN, B. C. (1969). *Trans. Met. Soc. AIME* **245**, 2089.
ALLEN, B. C. (1972). *Metall. Trans.* **3**, 2544.
AMATO, I., COLOMBO, R. L., and GRAPPIOLO, G. C. (1966). *Solid State Commun.* **4**, 237.
AUSTIN, A. E., RICHARD, N. A., and WOOD, V. E. (1966). *J. Appl. Phys.* **37**, 3650.
AYRAULT, G. and EHRLICH, G. (1972). *J. Chem. Phys.* **57**, 1788.
BAILEY, G. L. J., and WATKINS, H. C. (1950). *Proc. Phys. Soc.* **63B**, 350.
BARBOUR, J. P., CHARBONNIER, F. M., DOLAN, W. W., DYKE, W. P., MARTIN, E. E., and TROLAN, J. K. (1960). *Phys. Rev.* **117**, 1452.
BARNES, R. S., and MAZEY, D. J. (1963). *Proc. Roy. Soc. (London)* **275**, 47.
BARRER, R. M. (1941). "Diffusion in and through Solids." Cambridge Univ. Press, London and New York.
BARRER, R. M. (1967). *In* "The Solid-Gas Interface" (E. A. Flood, ed.), Vol. 2, p. 557. Dekker, New York.
BASSETT, D. W., and PARSLEY, M. J. (1970). *J. Phys. D: Appl. Phys.* **3**, 707.
BASSETT, G. A. (1958). *Phil. Mag.* **3**, 1042.
BASSETT, G. A. (1961). *In* "Electron Microscopy" (A. L. Howink and B. J. Spit, eds.), Vol. 1, p. 270. Vereniging voor Electron Microsc., Delft.
BASSETT, G. A. (1964). *In* "Condensation and Evaporation of Solids" (E. Rutner, P. Goldfinger, and J. P. Hirth, eds.), p. 599. Gordon and Breach, New York.
BAUER, E., and POPPA, H. (1972). *Thin Solid Films* **12**, 167.
BENARD, J. (1969). *Catal. Rev.* **3**, 93.
BETHGE, H. (1964). *Surface Sci.* **3**, 33.
BETHGE, H. (1969). *In* "Molecular Processes on Solid Surfaces" (E. Drauglis, R. D. Gretz, and R. I. Jaffee, eds.), pp. 569–585. McGraw-Hill, New York.
BETTLER, P. C., and CHARBONNIER, F. M. (1960). *Phys. Rev.* **119**, 85.
BLAKELY, J. M. (1963). *Prog. Mater. Sci.* **10**.
BLAKELY, J. M. (1966). *Acta Met.* **14**, 898.
BLAKELY, J. M., and JACKSON, K. A. (1962). *J. Chem. Phys.* **37**, 428.
BLAKELY, J. M., and LI, C. Y. (1966). *Acta Met.* **14**, 279.
BLAKELY, J. M., and MYKURA, H. (1961). *Acta Met.* **9**, 23.
BLAKELY, J. M., and MYKURA, H. (1962). *Acta Met.* **10**, 565.
BLAKELY, J. M., and OLSON, D. L. (1968). *J. Appl. Phys.* **39**, 3476.
BODDENBERG, B. (1970). *Surface Sci.* **22**, 39.
BODDENBERG, B., HAUL, R., and OPPERMANN, G. (1970). *Surface Sci.* **22**, 29.
BONZEL, H. P. (1970). *Surface Sci.* **21**, 45.
BONZEL, H. P. (1973). *In* "Structure and Properties of Metal Surfaces" (S. Shimodaira, ed.), Vol. 1. Maruzen, Tokyo.
BONZEL, H. P., and AARON, H. B. (1971). *Scripta Met.* **5**, 1057.
BONZEL, H. P., and GJOSTEIN, N. A. (1967). *Appl. Phys.Lett.* **10**, 258.
BONZEL, H. P., and GJOSTEIN, N. A. (1968a). *J. Appl. Phys.* **39**, 3480.
BONZEL, H. P., and GJOSTEIN, N. A. (1968b). *Phys. Status Solidi* **25**, 209.
BONZEL, H. P., and GJOSTEIN, N. A. (1969). *In* "Molecular Processes on Solid Surfaces" (E. Drauglis, R. G. Gretz, and R. I. Jaffee, eds.), pp. 533–568. McGraw-Hill, New York.
BONZEL, H. P., and GJOSTEIN, N. A. (1970). *Surface Sci.* **22**, 216.
BONZEL, H. P., and GJOSTEIN, N. A. (1972). *Phys. Status Solidi (a)* **14**, K59.
BONZEL, H. P., and KU, R. (1973). *J. Chem. Phys.* **59**, 1641.

BOWDEN, F. P., and SINGER, K. E. (1969). *Nature (London)* **222**, 977.

BRADSHAW, F. J., BRANDON, R. H., and WHEELER, G. (1964). *Acta Met.* **12**, 1057.

BURTON, J. J., and JURA, G. (1967a). *J. Chem. Phys.* **71**, 1937.

BURTON, J. J., and JURA, G. (1967b). *In* "Fundamentals of Gas Surface Interactions" (H. Saltsburg, J. N. Smith, and M. Rogers, eds.), p. 75. Academic Press, New York.

BURTON, W. K., CABRERA, N., and FRANK. F. C. (1951). *Phil. Trans. Roy. Soc. (London)* **243A**, 299.

CAHN, J. W. (1967). *J. Phys. Chem. Solids Suppl. 1* 681.

CARMAN, P. C. (1956). "Flow of Gases Through Porous Media." Butterworths, London and Washington, D.C.

CHAKRAVERTY, B. K. (1967). *J. Phys. Chem. Solids* **28**, 2401, 2413.

CHOI, J. Y., and SHEWMON, P. G. (1964). *Trans. Met. Soc. AIME* **230**, 123.

COLLINS, H. E., and SHEWMON, P. G. (1966). *Trans. Met. Soc. AIME* **236**, 1354.

DEBOER, J. H. (1968). "The Dynamical Character of Adsorption." Oxford Univ. Press (Clarendon), London and New York.

DELAMARE, F., and RHEAD, G. E. (1971). *Surface Sci.* **28**, 267.

DITTMAR, W., and MENNICKE, S. (1970). *Z. Phys. Chem. N.F.* **71**, 255.

DITTMAR, W., and NEUMANN, K. (1960). *Z. Elektrochem.* **64**, 297.

DRECHSLER, M. (1954). *Z. Elektrochem.* **58**, 327, 344, 340.

DREW, J. B., and AMAR, J. (1964). *J. Appl. Phys.* **35**, 533.

DREW, J. B., and PYE, J. J. (1963). *Trans. Met. Soc. AIME* **227**, 99.

DYKE, W. P., and BARBOUR, J. P. (1956). *J. Appl. Phys.* **27**, 356.

EHRLICH, G. (1966). *J. Chem. Phys.* **44**, 1050.

EHRLICH, G., and HUDDA, F. G. (1966). *J. Chem. Phys.* **44**, 1039.

EHRLICH, G., and KIRK, C. F. (1968). *J. Chem. Phys.* **48**, 1465.

ELLIS, W. P., and SCHWOEBEL, R. L. (1968). *Surface Sci.* **11**, 82.

FELLER, W. (1959). "An Introduction to Probability Theory and its Applications," p. 323. Wiley, New York.

FISHER, J. C. (1951). *J. Appl. Phys.* **22**, 74.

FRENKEL, J. (1924). *Z. Phys.* **26**, 117.

GAL, V. V. and BORISOV, V. T. (1971). *Fiz. Metal. Metalloved.* **31**, 973.

GAL, V. V., GRUZIN, P. L., and YUDINA, G. K. (1970). *Fiz. Metal. Metalloved.* **30**, 950.

GAL, V. V., GRUZIN, P. L., and YUDINA, G. K. (1973). *Phys. Status Solidi (a)* **15**, 659.

GEGUZIN, YA. E. (1969). *In* "Poverkhnostnayi Diffuziya i Rastekaniye" (Ya. E. Geguzin, ed.), pp.11–77. Izd. Nauke, Moscow.

GEGUZIN, YA. E., and KAGANOVSKY, YU. S. (1969). *In* "Poverkhnostnayi Diffuziya i Rastekaniye" (Ya. E. Geguzin, ed.), pp. 81–90. Izd. Nauke, Moscow.

GEGUZIN, YA. E., and KOVALEV, G. N. (1963a). *Fiz. Tverd. Tela* **5**, 1687.

GEGUZIN, YA. E., AND KOVALEV (1963b). *Dokl. Akad. Nauk (USSR)* **142**, 1290.

GEGUZIN, YA. E., KOVALEV, G. N., and RATNER, A. M. (1960). *Fiz. Metal. Metalloved.* **10**, 47.

GEGUZIN, YA. E., KOVALEV, G. N., and OVCHARENKO, N. N. (1964). *Fiz. Tverd. Tela* **5**, 3580.

GEGUZIN, YA. E., KAGANOVSKY, YU. S., and SLYOZOV, V. V. (1969). *J. Phys. Chem. Solids* **30**, 1173.

GEGUZIN, YA. E., KIBETS, V. I., and MAKAROVSKIY, N. A. (1972). *Fiz. Metal. Metalloved.* **33**, 800.

GEUSS, J. W. (1972). *Ned. Tijdschr. Vacuum Technol.* **10**, 59.

GILMER, G. H., and BENNEMA, P. (1972). *J. Appl. Phys.* **34**, 1347.

GIRIFALCO, L. A., and WEIZER, V. G. (1959). *Phys. Rev.* **114**, 687.

GJOSTEIN, N. A. (1961). *Trans. Met. Soc. AIME* **221**, 1039.

GJOSTEIN, N. A. (1963). *In* "Metal Surfaces: Structure, Energetics and Kinetics." Amer. Soc. Metals, Metals Park, Ohio.

GJOSTEIN, N. A. (1965). *In* "Adsorption et Croissance Cristalline." *Colloqu. Int. C.N.R.S. No. 152,* p. 97.

GJOSTEIN, N. A. (1967). *In* "Surfaces and Interfaces, I. Chemical and Physical Characteristics" (T. T. Burke, N. L. Reed and V. Weiss, eds.). Syracuse Univ. Press, Syracuse, New York.

GJOSTEIN, N. A. (1970a). *In* "Techniques of Metals Research" (R. F. Bunshah, ed.), Vol. IV, Part 2, Physicochemical Measurements in Metals Research (R. A. Rapp, ed.). Wiley (Interscience), New York.

GJOSTEIN, N. A. (1970b). *Metall. Trans.* **1**, 315.

GJOSTEIN, N. A. (1973). *In* "Diffusion." ASM Seminar Proc. Metals Park, Ohio.

GJOSTEIN, N. A., and BRAILSFORD, A. D. (1973). To be published.

GJOSTEIN, N. A., and HIRTH, J. P. (1965). *Acta Met.* **13**, 991.

GOODMAN, F. O. (1967). *Phys. Rev.* **164**, 1113.

GOMER, R. (1958). *J. Chem. Phys.* **28**, 457.

GOMER, R. (1959). *Discuss. Faraday Soc.* **28**, 23.

GRIMLEY, T. B. (1969). *In* "Molecular Processes on Solid Surfaces" (E. Drauglis, R. D. Gretz, and R. I. Jaffee, eds.) p. 299. McGraw-Hill, New York.

GULDEN, M. E. (1967). *J. Nucl. Mater.* **23**, 30.

GYULAI, Z. (1935). *Z. Krist.* (*A*) **91**, 149.

HAUL, R., and BODDENBERG, B. (1969). *Z. Phys. Chem. N.F.* **64**, 78.

HAYEK, K., and SCHWABE, U. (1972). *J. Vac. Sci. Technol.* **9**, 507.

HENRION, J. (1972). Thesis, Univ. of Paris.

HENRION, J., and RHEAD, G. E. (1970). *In* "Diffusion Processes." Gordon and Breach, New York.

HENNEY, J., and JONES, J. W. S. (1968). *J. Mater. Sci.* **3**, 158.

HENZLER, M. (1970a). *Surface Sci.* **19**, 159.

HENZLER, M. (1970b). *Surface Sci.* **22**, 12.

HERRING, C. (1951). *In* "Physics of Powder Metallurgy" (W. E. Kingston, ed.). McGraw-Hill, New York.

HILL, T. L. (1960). "Introduction to Statistical Thermodynamics." Addison-Wesley, Reading, Massachusetts.

HIRANO, K., and TANAKA, R. (1970). *Nippon Kinzoku Gakkai Kaiho* **9** (6), 341.

HIRTH, J. P. (1970). Private correspondence.

HIRTH, J. P., and POUND, G. M. (1963). *Progr. Mater. Sci.* **11**.

HOEHNE, K., and SIZMANN, R. (1971). *Phys. Status Solidi* (*a*) **5**, 577.

HRUSKA, S. J., and HIRTH, J. P. (1961). *Z. Elektrochem.* **65**, 479.

IKEDA, T. (1965). *J. Phys. Soc. Japan* **20**, 1259.

JOHNSON, D. L. (1969). *J. Appl. Phys.* **40**, 192.

JOST, W. (1952). "Diffusion in Solids, Liquids, Gases." Academic Press, New York.

KAGANOVSKIY, YU. S., and RATINOV, G. S. (1971). *Fiz. Metal. Metalloved.* **31**, 143.

KING, R. T., and MULLINS, W. W. (1962). *Acta Met.* **10**, 601.

KOSENKO, V. E., and KHOMENKO, L. A. (1961). *Fiz. Tverd. Tela* **3**, 2967.

KOSSEL, W. (1927). *Nachr. Ges. Wiss. Göttingen, Math.-Phys. KL.* 135.

KUCZYNSKI, G. C. (1949). *Trans. Met. Soc. AIME* **185**, 796.

KUCZYNSKI, G. C. (1961). "Powder Metallurgy," p. 11. Wiley (Interscience), New York.

LANG, B., JOYNER, R. W., and SOMORJAI, G. A. (1972). *Surface Sci.* **30**, 440, 454.

LANG, N. D., and KOHN, W. (1970). *Phys. Rev.* **B1**, 4555.

LANG, N. D., and KOHN, W. (1971). *Phys. Rev.* **B3**, 1215.

LANGMUIR, I., and TAYLOR, J. B. (1932). *Phys. Rev.* **40**, 463.

LENNARD–JONES, J. E. (1932). *Trans Faraday Soc.* **28**, 333.

LENNARD–JONÉS, J. E. (1937). *Proc. Phys. Soc.* (*London*) **49** (Extra Part), 140.

LEWIS, R., and GOMER, R. (1967). *Suppl. Nuovo Cim.* (*Serie I*) **5**, 506.

LEWIS, R., and GOMER, R. (1968). *Surface Sci.* **12**, 157.

LEWIS, R., and GOMER, R. (1969). *Surface Sci.* **17**, 333.

LI, S. Y., BLAKELY, J. M., and FEINGOLD, A. H. (1966). *Acta Met.* **14**, 1397.
LIFSHITZ, I. M., and SLEZHOV, V. V. (1961). *J. Phys. Chem. Solids* **19**, 35.
MAIYA, P. S., and BLAKELY, J. M. (1967). *J. Appl. Phys.* **38**, 698.
MAIYA, P. S., and ROUTBORT, J. L. (1972). *J. Mater. Sci.* **7**, 609.
MARINOVA, T. S., and ZUBENKO, YU. V. (1970). *Fiz. Tverd. Tela* **12**, 520.
MASSON, A., and KERN, R. (1968). *J. Crystal Growth* **2**, 227.
MASSON, A., METOIS, J. J., and KERN, R. (1970). *C.R. Acad. Sci. Paris.* **271**, 235, 398.
MASSON, A., METOIS, J. J., and KERN, R. (1971). *Surface Sci.* **27**, 463, 483.
MCALLISTER, P. V., and CUTLER, J. B. (1970). *Metall. Trans.* **1**, 313.
MCALLISTER, P. V., and CUTLER, I. B. (1972). *Metall. Trans.* **3**, 1673.
MCLEAN, M., and HIRTH, J. P. (1969). *Acta Met.* **17**, 237.
MELMED, A. J., and GOMER, R. (1961). *J. Chem. Phys.* **34**, 1802.
MELMED, A. J. (1965). *J. Chem. Phys.* **43**, 3057.
MENNICKE, S., and DITTMAR, W. (1972). *Z. Phys. Chem. N.F.* **79**, 10.
METOIS, J. J., GAUCH, M., MASSON, A., and KERN, R. (1972). *Surface Sci.* **30**, 43.
MEYER, H. J. (1968). *Nova Acta Leopoldina Suppl. 1* **34**, 155.
MILLS, B., and DOUGLAS, P. (1970). *Cent. Nat. Rech. Sci. Coll. Int. No. 187*, 207.
MILLS, B., DOUGLAS, P., and LEAK, G. M. (1969). *Trans. Met. Soc. AIME* **245**, 1291.
MÜLLER, E. W. (1949). *Z. Phys.* **126**, 642.
MÜLLER, E. W. (1957). *Z. Electrochem.* **61**, 43.
MULLINS, W. W. (1957). *J. Appl. Phys.* **28**, 333.
MULLINS, W. W. (1959). *J. Appl. Phys.* **30**, 77.
MULLINS, W. W. (1961). *Phil. Mag.* **6**, 1313.
MULLINS, W. W. (1963). *In* "Metal Surfaces: Structure, Energetics and Kinetics," p. 17. Amer. Soc. Metals, Metals Park, Ohio.
MULLINS, W. W., and SHEWMON, P. G. (1959). *Acta Met.* **7**, 163.
NEUMANN, G., and HIRSCHWALD, W. (1972). *Z. Phys. Chem. N.F.* **81**, 163.
NEUMANN, G., and NEUMANN, G. M. (1972). Diffusion Monograph Series (F. H. Wohlbier, ed.). Diffusion Inform. Center, Bay Village, Ohio.
NEUMANN, G. M., HIRSCHWALD, W., and STRANSKI, I. N. (1966). *Z. Natur.* **11a**, 807.
NEUMANN, K. (1938). *Z. Elektrochem.* **44**, 474.
NICHOLAS, M. G., and HODKIN, E. N. (1971). *J. Nucl. Mater.* **38**, 234.
NICHOLS, F. A. (1966). *J. Appl. Phys.* **37**, 2805.
NICHOLS, F. A., and MULLINS, W. W. (1965a). *J. Appl. Phys.* **36**, 1826.
NICHOLS, F. A., and MULLINS, W. W. (1965b). *Trans. Met. Soc. AIME* **233**, 1840.
NICKERSON, R. A., and PARKER, E. R. (1950). *Trans. Met. Soc. AIME* **42**, 376.
OLSON, D. L., PATIL, H. R., and BLAKELY, J. M. (1972). *Scripta Met.* **6**, 229.
PALYNKH, B. M., SIVERS, L. L., and NAZARENKO, L. F. (1971). *Fiz. Metal. Metallov.* **32**, 775.
PASHLEY, D. W., STOWELL, M. J., JACOBS, M. H., and LAW, T. J. (1964). *Phil. Mag.* **10**, 127.
PERDEREAU, J., and RHEAD, G. E. (1967). *Surface Sci.* **7**, 175.
PERDEREAU, J., and RHEAD, G. E. (1969). *In* "Colloque International sur la Structure et les Propriétés des Surfaces des Solides." Editions du CNRS, Paris.
PERDEREAU, J., and RHEAD, G. E. (1971). *Surface Sci.* **24**, 555.
PICHAUD, M., and DRECHSLER, M. (1972). *Surface Sci.* **32**, 341.
PICHAUD, M., and DRECHSLER, M. (1973). *Surface Sci.* **36**, 813.
PINES, B. YA., GREBENNIK, I. P., and GEKTINA, I. V. (1967). *Kristallogr.* **12**, 639.
PINES, B. YA., GREBENNIK, I. P., and ZYMAN, Z. Z. (1969). *Fiz. Metal. Metalloved.* **27**, 307.
PINES, B. YA., GREBENNIK, I. P., and GEKTINA, I. V. (1970). *Fiz. Metal. Metalloved.* **30**, 1320.
PYE, J. J., and DREW, J. B. (1964). *Trans. Met. Soc. AIME* **230**, 1500.
REYNOLDS, G. L. (1967). *J. Nucl. Mater.* **24**, 69.

RHEAD, G. E. (1969). *Surface Sci.* **15**, 353.

RHEAD, G. E. (1972). *Scripta Met.* **6**, 47.

RHEAD, G. E., and MYKURA, H. (1962). *Acta Met.* **10**, 843.

RHODIN, T. N., PALMBERG, P. W., and TODD, C. J. (1969). *In* "Molecular Processes on Solid Surfaces" (E. Drauglis, R. D. Gretz, and R. I. Jaffee, eds.), p. 499. McGraw-Hill, New York.

ROBERTSON, W. M. (1967). *In* "Sintering and Related Phenomena" (G. C. Kuczynski, N. A. Hooton, and C. F. Gibbon, eds.), p. 251. Gordon and Breach, New York.

ROBERTSON, W. M. (1969). *J. Nucl. Mater.* **30**, 36.

ROBERTSON, W. M., and CHANG, R. (1966). *Mater. Sci. Res.* **3**, 49.

ROSENBERG, R. (1972). *J. Vac. Sci. Technol.* **9**, 263.

ROULET, C. A. (1973). *Surface Sci.* **36**, 295.

ROULET, C. A., and BOREL, J. P. (1972). *C.R. Acad. Sci. Paris* **C274**, 2133.

RUSANOV, A. I. (1971). *In* "Progress in Surface and Membrane Science" (J. F. Danielli, M. D. Rosenberg, and D. A. Cadenhead, eds.), Vol. 4, pp. 57–114. Academic Press, New York.

RUTH, V., and HIRTH, J. P. (1964). *J. Chem. Phys.* **41**, 3139.

SCHRIEFER, J. R. (1972). *J. Vac. Sci. Technol.* **9**, 561.

SEARS, G. W. (1951). *J. Chem. Phys.* **25**, 637.

SEARS, G. W. (1953). *Acta Met.* **1**, 457.

SEARS, G. W. (1955). *Acta Met.* **3**, 361, 367.

SEARS, G. W. (1956). *Acta Met.* **4**, 268.

SHACKELFORD, J. F., and SCOTT, W. D. (1968). *J. Amer. Ceram. Soc.* **51**, 688.

SHEWMON, P. G. (1963). *J. Appl. Phys.* **34**, 755.

SHEWMON, P. G. (1964). *Trans. Met. Soc. AIME* **230**, 1134.

SHEWMON, P. G., and CHOI, J. Y. (1962). *Trans. Met. Soc. AIME* **224**, 589.

SHEWMON, P. G., and CHOI, J. Y. (1963). *Trans. Met. Soc. AIME* **227**, 515.

SHEWMON, P. G., and CHOI, J. Y. (1964). *Trans. Met. Soc. AIME* **230**, 449.

SICKAFUS, E. N., and BONZEL, H. P. (1971). *In* "Progress in Surface and Membrane Science" (J. F. Danielli, M. D. Rosenberg, and D. A. Cadenhead, eds.), Vol. 4, pp. 116–230. Academic Press, New York.

SINGER, K. W. (1970). *Cent. Rech. Sci. Colloq. Int. No. 187*, 199.

SOKOLSKAYA, I. L. (1956). *J. Tech. Phys. (USSR)* **26**, 1177.

SOMORJAI, G. A. (1972). "Principles of Surface Chemistry." Prentice-Hall, Englewood Cliffs, New Jersey.

SRINIVASAN, S. R., and TRIVEDI, R. (1973). *Acta.Met.* **21**, 611.

STRANSKI, I. N. (1928). *Z. Phys. Chem.* **136**, 259.

STRANSKI, I. N., and SUHRMANN, R. (1949). *Ann. Phys.* **1**, 153.

SUMNER, G. G. (1965). *Phil. Mag.* **12**, 767.

SUZOKA, T. (1965). *J. Phys. Soc. Japan* **20**, 1259.

TAN, Y. T., PERKETT, W., and POWELL, R. (1971). *J. Appl. Phys.* **42**, 4752.

TROFINOV, V. I., CHALYKH, A. E., and LUKYANOVICH, V. M. (1969). *Fiz. Tverd. Tela* **11**, 846.

TSONG, T. T. (1972). *Phys. Rev.* **B6**, 417.

VINEYARD, G. H. (1957). *J. Phys. Chem. Solids* **3**, 121.

VLADIMIROV, G. G., and SOKOLSKAYA, I. L. (1970). *Fiz. Tverd. Tela* **12**, 1553.

VLADIMIROV, G. G., MEDVEDEV, B. K., and SOKOLSKAYA, I. L. (1970a). *Fiz. Tverd. Tela* **12**, 539.

VLADIMIROV, G. G., MEDVEDEV, B. K., and SOKOLSKAYA, I. L. (1970b). *Fiz. Tverd. Tela* **12**, 1423.

VOLMER, M. (1939). "Kinetik der Phasenbildung." Verlag Th. Steinkopff, Dresden.

VOLMER, M., and ESTERMANN, I. (1921). *Z. Phys.* **7**, 1.

WAGNER, C. (1961). *Z. Elektrochem.* **65**, 581.

WHIPPLE, R. T. P. (1954). *Phil. Mag.* **45**, 1225.

WILLERTZ, L. E., and SHEWMON, P. G. (1970). *Metall. Trans.* **1**, 2217.

WILLIAMS, F. L., and BOUDART, M. (1973). *J. Catal.* **30**, 438.
WINEGARD, W. C., and CHALMERS, B. (1952). *Can. J. Phys.* **30**, 422.
WINTERBOTTOM, W. L. (1969). *J. Appl. Phys.* **40**, 3803.
WOLFE, J. R., and WEART, H. W. (1969). *In* "The Structure and Chemistry of Solid Surfaces"
(G. A. Somorjai, ed.). Wiley, New York.
WYNBLATT, P. (1969). *Phys. Status Solidi* **36**, 797.
WYNBLATT, P., and GJOSTEIN, N. A. (1968). *Surface Sci.* **12**, 109.
WYNBLATT, P., and GJOSTEIN, N. A. (1970). *Surface Sci.* **22**, 125.
WYNBLATT, P., and GJOSTEIN, N. A. (1974). *Progr. Solid State Chem.* **9**, (to appear).
ZAHN, R. (1964). Thesis, Jernkontorets Lab. for Powder Metallurgy, Stockholm.

7

Interaction of Atoms and Molecules with Surfaces

J. W. GADZUK

NATIONAL BUREAU OF STANDARDS
WASHINGTON, D.C.

I. INTRODUCTION

Within the past few years the importance of the chemical state of solid surfaces has become recognized by solid state physicists. The role of the surface in such areas as catalysis, corrosion, and microminiaturized solid state circuitry is obvious. On the other hand it is only recently that the extreme surface sensitivity of experimental techniques involving a charged particle as either the exciting or excited particle, such as uv photoemission, ESCA, and ion neutralization spectroscopy, has been realized. In order that these techniques might be used to gather information on bulk properties, one must understand the surface properties of the sample and then be able to relate these properties to the bulk properties. Appelbaum (Volume I) has discussed the current theories of the electronic properties of clean surfaces. These properties can be drastically altered by exposing the clean surface to a gaseous contaminant in which the atoms or molecules adsorb on the surface. The type of adsorption is broadly divided into two different classes: physisorption, in which the heat of adsorption is typically much smaller than 1 eV; and chemisorption, in which the heat of adsorption is 1 eV or more. The noble gases are examples of atomic species which physisorb via weak van der Waals interactions. Simple molecules such as H_2 fall into both categories, depending upon the substrate material as well as the crystallographic orientation of the surface. In general if a stable molecule is dissociated by the molecule–surface interaction, the resulting atomic products will then form strong chemical bonds with the substrate. On the other hand, if the molecule is not dissociatively adsorbed, the resulting bond is much weaker and is principally due to polarization forces. In this chapter attention will be directed mainly to the low coverage chemisorption limit of adsorption, where interactions between adatoms are much smaller than the adatom–substrate interaction.

At least three different philosophical approaches to a chemisorption theory are currently being followed. The first, which might be termed the semiempirical theory, is a logical extension of the Pauling theory of the chemical bond (Pauling, 1960) and has been followed by several groups (Gyftopoulos and Levine, 1962a, b; Flaim and Ownby, 1972; Weinberg and Merrill, 1972), producing moderate success in correlating chemisorption data such as work function changes and chemisorption energies with individual adatom and substrate properties. We will not discuss this approach. Although in its infancy, the second possibility might be called a first principles theory, which is the chemisorption equivalent of an APW energy band calculation. Using the x–α method (Slater and Johnson, 1972), this line is being pioneered by Messmer et al. (1974). Due to the large amount of computation required for such a calculation, it should be quite a time before meaningful results of such theories are readily accessible to the nonspecialist trying to interpret experimental data.

The final tack, which we will concentrate on in this chapter, is referred to as the model Hamiltonian approach. The steps in the implementation of such a theory are the following. From physical considerations the dominating mechanism in the bond formation is assumed, i.e., Heitler–London correlations, molecular orbital (MO) formation, or dielectric screening. A relatively simple Hamiltonian for the coupled adsorbate–substrate system, emphasizing the particular bond mechanism, is then written [such as the Anderson (1961), Hubbard (1963), or finite jellium (Gadzuk, 1969a; Smith *et al.*, 1973)]. Due to the simplicity of the assumed Hamiltonian, diagonalization (perhaps after linearizing) is then possible, resulting in exact solutions to model problems. The solutions are given in terms of two or three quantities, i.e., hopping matrix elements, overlap integrals, and screened intraatomic Coulomb integrals, which in principle are calculable but in practice are usually treated as parameters. The major advantages of this approach are that a physically realistic model incorporating the principal interaction effects is easily accessible, and the model Hamiltonians should predict correct systematic trends as either substrates or adsorbates are varied. Since the model Hamiltonians focus attention on the interaction forces, it is possible to understand, on a microscopic level, what is "happening" in chemisorption. For these reasons we will consider only the model Hamiltonian approach in this chapter, which should complement and/or update several reviews (Horiuti and Toya, 1969; Wojciechowski, 1971; Grimley, 1971a; Schrieffer, 1972; Lang, 1973).

The chapter is structured as follows. In Section II a model Hamiltonian involving atomic orbitals, the Anderson model, and its principal features will be presented. The method of group orbitals in chemisorption problems, developed by Grimley (1971b, 1972, 1973) and Thorpe (1972), will be presented in Section III. This method is quite useful in clarifying the relationship between the localized molecular orbital bonds of chemistry and the delocalized bands of solid state physics. Furthermore, the crystallographic aspects of the problem are necessarily included in a straightforward way. The surface molecule limit of group orbital theory is then discussed. A treatment of chemisorption as the dielectric response of the electron gas to a charged impurity, as has been performed for both hydrogen and alkali atom chemisorption, is given in Section IV. Specific applications of chemisorption theory to systems that have been experimentally studied in detail are given in Section V. In particular hydrogen on W, alkali atoms on transition metals, and group V transition metals on W are considered.

II. MODEL HAMILTONIAN

Several chemisorption studies (Grimley, 1971a, b, 1972, 1973; Schrieffer, 1972; Thorpe, 1972; Bennett and Falicov, 1966; Gadzuk, 1967a, 1969; Edwards and Newns, 1967; Newns, 1969a; Gadzuk *et al.*, 1971; Muscat and Newns, 1972) have been based on a picture in which the interacting adatom–metal system is describable within the Anderson model of localized real and virtual impurity states (Anderson, 1961; Schrieffer and

Mattis, 1965; Kjollerström *et al.*, 1966; Heeger, 1969). Within this model, states of the interacting metal–atom system are constructed from linear combinations of the unperturbed continuum of metal states $|\mathbf{k};\sigma\rangle$ with eigenvalues $\varepsilon_{\mathbf{k},\sigma}$ and the discrete ground state of the atom $|a;\sigma\rangle$ with eigenvalue ε_{a0}. The label $\mathbf{k}$ represents the dynamical quantum numbers of the metal states, usually wave number, and σ is the spin of the electron in the given state. When spin is unimportant, this index will often be omitted. Excited states of the atom are neglected. The coupling between the adatom and the metal is represented by the off-diagonal matrix element $V_{ak} = \langle a|H|\mathbf{k}\rangle$, in which H is the full Hamiltonian of the combined system. The major feature of the Anderson model, which makes exact solution difficult, is the allowance for electron–electron interactions between electrons of opposite spin when they are both present on the adatom. This repulsive interaction tends to raise the energy ε_{a0} of the atomic state. Thus there is an energetic difference between a state with a single spin on the adatom (magnetic) and a state with the same net charge but with both spin-up and spin-down electrons present (possibly nonmagnetic). It is this effect together with the effect of V_{ak} that gives rise to the possibility of energy lowering in the adsorbed state. Mathematically the model Hamiltonian is given, in second quantized notation, by

$$H = \sum_{\mathbf{k},\sigma} \varepsilon_{\mathbf{k}} n_{\mathbf{k},\sigma} + \sum_{\sigma} \varepsilon_{a0}\, n_{a\sigma} + \sum_{\mathbf{k},\sigma} (V_{ak}\, c_{a,\sigma}^{\dagger} c_{\mathbf{k},\sigma}$$
$$+ V_{ak}^{*}\, c_{\mathbf{k},\sigma}^{\dagger} c_{a,\sigma}) + U n_{a\sigma} n_{a-\sigma}. \tag{1}$$

The diagonal matrix elements of the number operators $n_{k\sigma}$, $n_{a\sigma}$ give the number of electrons in the state $|\mathbf{k};\sigma\rangle$ and $|a;\sigma\rangle$. The c operators are Fermion creation and annihilation operators. We will always work in the zero temperature limit in which the Fermi function is a step function at $\varepsilon = 0$, the Fermi energy. In Eq. (1) the first two terms are the unperturbed Hamiltonians of the metal and atom. The third term is the atom–metal coupling term that transfers electrons between the two entities. The last term in Eq. (1) is the intra-atomic Coulomb repulsion between up and down spin electrons on the adatom. The effective strength of this repulsion is given by the quantity U. In the original theory (Anderson, 1961), U was taken to be the calculated gas phase Coulomb integral

$$U = \int d^3 r_1\, d^3 r_2\, |\phi_{a\sigma}(\mathbf{r}_1)|^2 (e^2/r_{12})|\phi_{a-\sigma}(\mathbf{r}_2)|^2, \tag{2}$$

with $\phi_{a\sigma}(\mathbf{r}_1)$ the real space wave function for the abstract state $|a;\sigma\rangle$ and $r_{12} \equiv |\mathbf{r}_1 - \mathbf{r}_2|$. The values of U calculated from Eq. (2) are too large even for gas phase work, due to the fact that no correlation effects are built into the wave functions. In reality the electrons will attempt to stay apart so that they see a weaker repulsion. In atomic physics this correlation effect has been treated by variational procedures in which polarization corrections to the zero order wave functions are included (Bethe and Salpeter, 1957). In the realm of the Anderson model, Schrieffer and Mattis (1965) have included correlation effects within the t-matrix approximation and find that the uncorrelated U is reduced considerably. To see the possible errors in Eq. (2) when applied to H, for instance, straightforward calculations using 1s wave functions give $U \sim 17\,\mathrm{eV}$, implying that H^-, a hydrogen negative ion, is unstable. In point of fact the electron affinity $\equiv A$ of H is $\sim 0.7\,\mathrm{eV}$, so the relevant quantity representing the electron repulsion, including correlations, should be $U = V_i - A \simeq 12.9\,\mathrm{eV}$, where V_i is the ionization potential of H (Newns, 1969). Still another correction to Eq. (2), which is of paramount importance, is due to screening of e^2/r_{12} by the free conduction band electrons of the solid (Herring, 1966). The simplest screening correction, Fermi–Thomas screening, with the replacement

$$(e^2/r_{12}) \rightarrow (e^2/r_{12}) \exp(-k_{\mathrm{FT}}\langle r_{12} \rangle),$$

with k_{FT} the Fermi–Thomas screening parameter and $r_{12} \sim \langle r_{12} \rangle$ the average electron–electron separation, can reduce U to the range $\sim(0.1 - 0.5) \times U$, a nontrivial effect in even qualitative considerations.

The physical picture emerging from the Anderson model is that of quasi-localized virtual states. More specifically, as shown in Fig. 1a, the non-interacting atom and metal are characterized by energies previously discussed. The metal work function is ϕ and the occupied portion of the conduction band has a width E_F, the Fermi energy. The possibility of a narrow d band centered at energy ε_d (with respect to energy zero at the Fermi level) is indicated by the dense level spacing. As the atom is brought to the surface, the V_{ak} and U terms shift both the valence level ε_{a0} and the affinity level at A below the vacuum potential. Furthermore all metal states with energies "near" the energy of the shifted atom state mix well with the atomic state. Consequently, if a hole is created in the adatom virtual state [in an experiment such as field emission (Gadzuk and Plummer, 1973), photoemission (Eastman and Cashion, 1971; Waclawski and Plummer, 1973; Feuerbacher and Fitton, 1973), or ion neutralization spectroscopy (Hagstrum, 1972)], it will have a very short lifetime with respect to being filled by a metal electron in the energy range $\varepsilon_a \pm \Delta$. Thus it appears that the discrete adatom state has been converted into a broadened virtual state of width $\equiv 2\Delta_a$. Figure 1b depicts the case in which the virtual state is almost totally below the Fermi level and is thus occupied. Figure 1c shows the case in which the virtual level lies above the Fermi level and is thus mostly unoccupied.

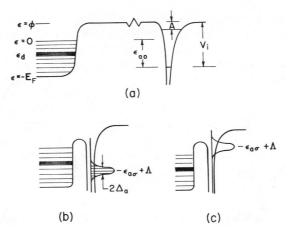

Fig. 1. (a) Schematic potential and energy level diagram for noninteracting atom and metal. The occupied portion of the conduction band lies within the range $-E_F \leqslant \varepsilon \leqslant 0$. A narrow d band is centered at $\varepsilon = \varepsilon_d$. (b) Adsorption for which the broadened atomic virtual state lies below the Fermi level and is thus totally occupied. (c) Ionic adsorption for which the broadened valence level lies above the Fermi level and is thus almost totally unoccupied.

The classic work in the development of the Anderson model for chemisorption is due to Newns (1969a). In his paper H chemisorption on transition metals is described with particular emphasis on the magnetic state of the adsorbed H atom. Equation (1) was solved in the unrestricted Hartree–Fock (HF) approximation [that is, $n_{a\sigma} n_{a-\sigma} \approx$ constant $+$ $(\langle n_{a\sigma} \rangle n_{a-\sigma} + \langle n_{a-\sigma} \rangle n_{a\sigma})$, with $\langle n_{a\sigma} \rangle$ the average value of the electron charge with spin σ on the adatom]. This procedure has been criticized because the HF approximation overemphasizes the tendency toward magnetic solutions. Newns points out that this is not a problem when dealing with nonmagnetic HF solutions. His analysis of H on several three-dimensional transition metal surfaces indicates that nonmagnetic behavior is expected and thus a HF solution should be reasonable.

Following Newns (1969a), the full Hamiltonian of Eq. (1) can be written as the sum of spin dependent Hamiltonians H^σ:

$$H = \sum_\sigma H^\sigma, \tag{3a}$$

with

$$H_\sigma = \varepsilon_{a\sigma} n_{a\sigma} + \sum_{\mathbf{k}} \varepsilon_{\mathbf{k}} n_{\mathbf{k}\sigma} + \sum_{\mathbf{k}} (V_{a\mathbf{k}} c_{a\sigma}^\dagger c_{\mathbf{k}\sigma} + V_{a\mathbf{k}}^* c_{\mathbf{k}\sigma}^\dagger c_{a\sigma}) \tag{3b}$$

and

$$\varepsilon_{a\sigma} = \varepsilon_{a0} + U_e \langle n_{a-\sigma} \rangle. \tag{3c}$$

The energy $\varepsilon_{a\sigma}$ defines an effective atomic energy level that allows for the fact that an electron of spin σ feels the repulsive interaction of the spin-σ electron while on the adatom. Here U has been replaced by the symbol

U_e, denoting the modified U due to correlations and/or screening effects. An adatom Green's function can be obtained from Eq. (3b) in the unperturbed representation

$$G_{aa}^\sigma(\varepsilon) = \left[\varepsilon - \varepsilon_{a\sigma} - \sum_{\mathbf{k}} \frac{|V_{a\mathbf{k}}|^2}{\varepsilon - \varepsilon_{\mathbf{k}} + i\delta} \right]^{-1} \equiv [\varepsilon - \varepsilon_{a\sigma} - \Lambda(\varepsilon) + i\Delta_a(\varepsilon)]^{-1}, \qquad (4a)$$

where

$$\Delta_a(\varepsilon) = -\,\mathrm{Im} \sum_{\mathbf{k}} |V_{a\mathbf{k}}|^2/(\varepsilon - \varepsilon_{\mathbf{k}} + i\delta) = \pi \sum_{\mathbf{k}} |V_{a\mathbf{k}}|^2\, \delta(\varepsilon - \varepsilon_{\mathbf{k}}) \qquad (4b)$$

and

$$\Lambda(\varepsilon) = \frac{P}{\pi} \int_{-\infty}^{+\infty} \frac{\Delta_a(\varepsilon')}{(\varepsilon - \varepsilon')}\, d\varepsilon', \qquad (4c)$$

with $\delta \to 0+$ and P denoting a principal part integral. The density of states of spin σ at the adatom is given by

$$\rho_{aa}^\sigma(\varepsilon) = -\,(1/\pi)\,\mathrm{Im}\, G_{aa}(\varepsilon) \qquad (5)$$

and the total electron charge number for electrons with spin σ by

$$\langle n_{a\sigma} \rangle = \int_{-\infty}^{0} \rho_{aa}^\sigma(\varepsilon)\, d\varepsilon. \qquad (6)$$

The energy of the adatom–metal level is given by the "poles" of the adatom Green's function, Eq. (4a),

$$\varepsilon - \varepsilon_{a\sigma} - \Lambda(\varepsilon) = 0. \qquad (7)$$

If this value of ε falls within the allowed electron energy band of the metal, then there are some $\varepsilon_{\mathbf{k}} = \varepsilon$. Consequently, from Eq. (4b), $\Delta_a(\varepsilon)$ is nonzero, in which case the adatom level can be thought of as a resonance or virtual state rather than a true eigenstate. The density of states from Eqs. (4a) and (5) is

$$\rho_{aa}^\sigma(\varepsilon) = \frac{1}{\pi}\, \frac{\Delta_a(\varepsilon)}{(\varepsilon - \varepsilon_{a\sigma} - \Lambda(\varepsilon))^2 + \Delta_a^2(\varepsilon)}. \qquad (8)$$

However, it is also possible for a truly localized impurity state to form if ε given by Eq. (7) falls outside the metal band (Koster and Slater, 1954; Clogston, 1962). When this occurs, the density of states and total charge per spin are given by

$$\rho_{aal}^\sigma(\varepsilon) = \langle n_{a\sigma} \rangle_l\, \delta(\varepsilon - \varepsilon_{a\sigma} - \Lambda(\varepsilon)), \qquad (9a)$$

and from Eqs. (4a) and (5),

$$\langle n_{a\sigma} \rangle_l = \left[1 - \frac{\partial \Lambda(\varepsilon)}{\partial \varepsilon} \Bigg|_{\varepsilon = \varepsilon_{a\sigma} + \Lambda(\varepsilon)} \right]^{-1}, \qquad (9b)$$

with subscript l representing localized and the derivative in Eq. (9b) evaluated at the energy of the localized state. Note that if V_{ak} in Eq. (4b) was replaced by some average value and thus taken outside the summation, then

$$\Delta_a \approx \pi \rho(\varepsilon) \langle |V_{ak}|^2 \rangle_{av}, \tag{10}$$

with $\rho(\varepsilon)$ the metal density of states at the energy ε given by Eq. (7). Under this assumption ρ_{aa}^σ given by Eq. (8) is simply a Lorentzian characterized by two parameters, the level center and width, and it is in this form that most spectroscopic data has been analyzed. This is not completely satisfactory, since many of the observed adatom densities of states are quite narrow, with Δ_a(Observed) $\lesssim 1$ eV. From Eq. (10) this implies that $|V_{ak}| \lesssim 0.75$ eV. On the other hand the strength of the chemisorption bond or the bond energy is typically $\gtrsim 3$ eV. Since the chemisorption energy, for strong localized chemical bonds, is directly determined by the magnitude of V_{ak} (Grimley, 1971a; Schrieffer, 1972; Newns, 1969a), the observed bond energies imply that $|V_{ak}| \gtrsim 3$ eV, which is in contradiction with the observed level widths. The resolution of this problem will be given in Section III.

The chemisorption energy ($\equiv E_{chem}$) is defined as the difference in energy between the initial and final states and with the present conventions is positive in sign for stability. It is given by

$$E_{chem} = E_{in} - \sum_\sigma \int_{-\infty}^0 \varepsilon \rho_{aa}^\sigma(\varepsilon) \, d\varepsilon + U_c \langle n_{a+} \rangle \langle n_{a-} \rangle,$$

where E_{in} is the initial energy of the electrons involved in the bond when $V_{ak} = 0$. The screened Coulomb interaction is subtracted from E_{fin} to compensate for double counting (Schrieffer and Mattis, 1965; Kjollerström et al., 1966).

The final point concerning the Anderson model is the requirement of self-consistency. From Eqs. (3c), (6), and (8) it is apparent that the population of spin σ electrons $\langle n_{a\sigma} \rangle = \langle n_{a\sigma}(n_{a-\sigma}) \rangle$; that is, it is a function of the occupation of spin-σ electrons and vice versa and thus, from Eq. (3c), so are the adatom energies. This condition generates two equations that must be satisfied simultaneously. If it is assumed that Λ and Δ_a do not vary with energy, then Eqs. (3c), (6), and (8) yield

$$\langle n_{a\pm\sigma} \rangle = \frac{1}{\pi} \cot^{-1} \left(\frac{\varepsilon_{a0} + U_c \langle n_{a\mp\sigma} \rangle + \Lambda}{\Delta_a} \right). \tag{11a,b}$$

Equations (11a) and (11b) always have a solution in which $\langle n_{a+\sigma} \rangle = \langle n_{a-\sigma} \rangle$, the nonmagnetic solution. However, if solutions exist for which $\langle n_{a+\sigma} \rangle \neq \langle n_{a-\sigma} \rangle$, then the resulting electronic configuration on the adatom possesses a net magnetic moment, and this state is energetically favored over the nonmagnetic state. It must be recalled that the reduction of the Anderson Hamiltonian has been done within HF, which overemphasizes the tendency toward magnetism. Thus if Eqs. (11a) and (11b) yield magnetic solutions, a better approximation than HF is required. Henceforth, considering only nonmagnetic solutions, the self-consistency condition on the total adatom electron charge is

$$\langle n_a \rangle = \frac{2}{\pi} \cot^{-1} \left(\frac{\varepsilon_{a0} + \frac{1}{2} U_e \langle n_a \rangle + \Lambda}{\Delta_a} \right), \tag{12}$$

where $\langle n_a \rangle \equiv \langle n_{a+\sigma} \rangle + \langle n_{a-\sigma} \rangle$ is the total charge. From here on, spin indices will be omitted unless otherwise noted.

The Anderson model is basically a molecular orbital (MO) theory in contrast to the Heitler–London-based theories of Schrieffer and co-workers (Schrieffer and Gomer, 1971; Einstein and Schrieffer, 1973; Paulson and Schrieffer, 1975). The easily visualized bond mechanisms in an MO theory make it attractive to work with. In spite of its limitations, it can provide useful insight into chemisorption mechanisms. As Heine has recently pointed out, the electron state of a solid can be considered to be a giant molecular orbital. "As we pull the atoms of a small molecule apart, the wave function changes continuously from one extreme where a molecular orbital description is a good approximation, to the other extreme of a Heitler–London wave function. In solids the two regions are divided by a phase transition, the Mott transition, and most commonly studied solids lie on the molecular orbital or Bloch side of the line" (Heine, 1972). With this in mind, it is not unreasonable to suppose that the electronic structure of an adatom on the surface can be understood within a molecular orbital theory. What cannot be understood, in a straightforward way, is the binding energy, since this requires a knowledge of the potential energy curve on both sides of the Mott transition and the molecular orbital theory breaks down for the insulator phase. As long as one is interested only in the properties on the metal side of the Mott transition, which could be the case in the chemisorbed state of small atoms, then the molecular orbital theories should be a valid vehicle of expression. Due to space limitations, we will direct attention to the MO rather than valence bond theories.

III. GROUP ORBITALS

A. General Properties

Considerable insight into the nature of bonding effects in solids and at surfaces has been gained through chemical theories that emphasize the directional properties of atomic orbitals on the individual atoms of the solid (Weinberg and Merrill, 1972; Goodenough, 1963; Tamm and Schmidt, 1969; Knor, 1971; Phillips, 1969). The major problem with the localized bond picture is how to make such theories consistent with the band theory of solids. Although an electron wave function of the solid possesses directional properties that can be put into correspondence with the properties of atomic orbitals, it is not possible to specify what energy is associated with such atomic orbitals to a degree more accurate than somewhere within the band or bands derived from the particular atomic orbital. For the transition metals of interest in chemisorption studies, the s-band widths are of order 10 eV and the d-band widths are as large as 5 eV. This degree of uncertainty in orbital energies has hindered the development of quantitative theories of chemisorption based on the orbital picture. The group orbital techniques utilized by Grimley (1971a, b, 1972, 1973) and Thorpe (1972) and outlined in this section offer a means of simplifying the description of localized chemical bonds in systems in which metal band effects might at first seem to vitiate the idea of localized bonds. Briefly stated, a description is sought in which molecular orbitals are formed from the valence atomic orbitals of the adsorbate and a sum of localized orbitals centered on a small set of substrate atoms. The choice of the particular set of substrate orbitals and centers is dictated by the local symmetry of the bond configuration. Calculations are usually performed within the two-centered–tight-binding approximation.

The physical content of the group orbital, bond–band picture can be demonstrated with the following model. Consider a simple cubic p band metal with $2a \equiv$ lattice constant and z the direction normal to the surface. Next suppose that an adsorbate with either an s or p valence electron adsorbs interstitially in the bridge site between two surface atoms as shown in Fig. 2. In the figure, p orbitals centered on the lattice sites of the substrate are drawn and the alternative adsorbate orbitals are also shown. Although the orbital diagram suggests a simple interpretation, namely that the adsorbate overlaps most strongly with the p_x orbitals of the substrate, the band nature of the solid must still be dealt with. From Eqs. (4) and (5), the object of immediate importance is the matrix element V_{ak}. This is a measure of the coupling strength between the adsorbate state $|a\rangle$ and the substrate eigenstates $|k\rangle$ which can be expressed within a localized representation as

$$|k\rangle = 1/\sqrt{N} \sum_i e^{i k \cdot R_i} \phi(r - R_i), \qquad (13)$$

where N is the number of atoms in the crystal, R_i is the position of the ith ion core, and $\phi(r - R_i)$ is an atomic-like orbital centered on the ith ion core. For a realistic, three-dimensional surface calculation, $|k\rangle$ would be modified to reflect the surface boundary conditions. The most important

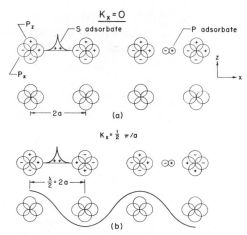

Fig. 2. (a) p band substrate orbitals (for $\mathbf{k} = 0$) centered on each lattice site. Also shown is the wave function for an interstitial s and p adsorbate. (b) Similar situation but with $k_x = \frac{1}{2}\pi/a$. The sinusoidal function is also shown.

correction would be the substitution $e^{i\mathbf{k}\cdot\mathbf{R}_i} = e^{i\mathbf{k}_{\parallel}\cdot\mathbf{R}_i} f(k_z, \mathbf{R}_i)$, where $\mathbf{k}_{\parallel}(k_z)$ is the transverse (normal) component of $\mathbf{k}$ and $f(k_z, \mathbf{R}_i)$ is a simple function including surface effects. As pictured in Fig. 2, the p orbital–adsorbate matrix element with either nearest neighbor substrate orbital is of the same sign, and the sum of the matrix elements is nonvanishing, a bonding situation. This can be thought of as constructive interference. On the other hand, the matrix elements between the adsorbate s orbital and the substrate nearest neighbors are of opposite sign, and the sum vanishes due to the relative phases of the substrate and adsorbate orbitals, a result of destructive interference. Comparing the substrate orbitals shown in Fig. 2 with the Bloch function given by Eq. (13), it is noted that the orbitals pictured are of the form of the $\mathbf{k} = 0$ Bloch function. The Bloch states are the periodic atomic orbitals modulated by the plane wave function. For Bloch states in which one half a wavelength equals the ion core separation, $\lambda/2 = 2a$ or $k_x = \frac{1}{2}\pi/a$, the phases of alternate substrate orbitals are reversed. When this occurs, the s orbital couples to the p band whereas the p orbital does not. The variation of relative orbital phases with $\mathbf{k}$ is the basic physical mechanism providing for the selectivity with $\mathbf{k}$ and thus energy $\varepsilon_{\mathbf{k}}$ for adsorbate orbital mixing with substrate orbitals. It is this effect that dictates that $V_{a\mathbf{k}}$ in Eq. (4b) must remain within the summation and thus gives $\Delta_a(\varepsilon)$ a nontrivial energy dependence.

   To quantify these physical notions, follow Grimley and suppose that the $V_{a\mathbf{k}}$ coupling, in the surface molecule limit, can be decomposed into adatom coupling with the substrate orbitals of a certain symmetry centered on the adatom nearest neighbors. This group orbital will be labeled

$$|g\rangle = 1/\sqrt{N_g} \sum_i a_i |\phi_g(\mathbf{R}_i)\rangle, \tag{14}$$

with $a_i = \pm 1$ depending upon the site and orbital symmetry, N_g the number of centers in the orbital, $\mathbf{R}_i$ the location of the ith atom, and ϕ_g the substrate orbital of g symmetry. Next it is assumed that the total Δ can be decomposed into a noninterfering sum of terms for each group orbital

$$\Delta(\varepsilon) \simeq \sum_g \Delta_g(\varepsilon), \tag{15}$$

and usually one g in the sum will dominate. As a result of the directional properties of the group orbitals, it is imagined that as $|a\rangle$ couples with $|\mathbf{k}\rangle$ in the matrix elements of Eq. (4b), it picks out of $|\mathbf{k}\rangle$ only that part projecting on to $|g\rangle$. Thus the level width is approximated by

$$\Delta_g(\varepsilon) \simeq \pi \sum_{\mathbf{k}} |\langle a|H|g\rangle \langle g|\mathbf{k}\rangle|^2 \, \delta(\varepsilon - \varepsilon_g(\mathbf{k})), \tag{16}$$

with $\varepsilon_g(\mathbf{k})$ the diagonal components of the metal Hamiltonian between states similar to those given by Eq. (13) but with $\phi(\mathbf{R}_i)$ replaced by $\phi_g(\mathbf{R}_i)$. Inserting Eq. (14) into the matrix element of Eq. (16) yields

$$\langle a|H|g\rangle \langle g|\mathbf{k}\rangle = 1/N_g \sum_{i,j} a_i a_j \langle a|H|\phi_g(\mathbf{R}_i)\rangle \langle \phi_g(\mathbf{R}_j)|\mathbf{k}\rangle. \tag{17}$$

The a factors are such that the product $a_i \langle a|H|\phi_g(\mathbf{R}_i)\rangle$ is constant for all values of i, and this quantity is also independent of $\mathbf{k}$. This matrix element involves the overlap of atomic orbitals on the adsorbate with one of the N_g equivalent substrate atoms, and its calculation is a problem mainly of molecular chemistry. The following expression defines β_g:

$$\sum_i a_i \langle a|H|\phi_g(\mathbf{R}_i)\rangle \equiv N_g \beta_g. \tag{18}$$

The solid state physics aspect of the problem is contained in the overlap integral of Eq. (17),

$$\langle g|\mathbf{k}\rangle = 1/\sqrt{N} \sum_j \frac{a_j}{N_g} e^{i\mathbf{k}\cdot\mathbf{R}_j}, \tag{19}$$

where the tight-binding expression for the Bloch functions, Eq. (13), has been used and $\langle \phi_g(\mathbf{R}_i)|\phi_g(\mathbf{R}_j)\rangle = \delta_{i,j}$. Equation (19) reduces to an explicit trigonometric function once a definite bond configuration is assumed.

Combining Eqs. (16)–(19), the level width function becomes

$$\Delta_g(\varepsilon) = \pi |N_g \beta_g|^2 \sum_{\mathbf{k}} \frac{1}{N} \left| \sum_j \frac{a_j}{N_g} e^{i\mathbf{k}\cdot\mathbf{R}_j} \right|^2 \delta(\varepsilon - \varepsilon_g(\mathbf{k})) \equiv \pi |N_g \beta_g|^2 \rho_g(\varepsilon), \qquad (20)$$

which defines $\rho_g(\varepsilon)$, the group orbital virtual resonance. The utility of the group orbital approach is the natural separation into a molecular orbital portion dictating the strength of the interaction $N_g \beta_g$ and a solid state portion $\rho_g(\varepsilon)$, which allows for the adsorbate coupling to states in a preferential region of $\mathbf{k}$ space. It is this preferential coupling that controls the assignment of effective energies to the substrate orbitals.

To be more specific, consider the cases shown in Fig. 2. The p(s) adsorbate couples to the group orbital $|g_{p(s)}\rangle = (1/\sqrt{2})(\phi_g(\mathbf{r} - \mathbf{R}_1) + (-)\phi_g(\mathbf{r} - \mathbf{R}_2))$, and thus for the p adsorbate, $a_1 = a_2 = 1$, whereas for the s adsorbate, $a_1 = -a_2 = 1$. Consequently, from Eqs. (19) and (20),

$$\rho_p(\varepsilon) = \sum_{\mathbf{k}} (1/N) \cos^2 k_x a \, \delta(\varepsilon - \varepsilon_g(\mathbf{k})) \qquad (21a)$$

$$\rho_s(\varepsilon) = \sum_{\mathbf{k}} (1/N) \sin^2 k_x a \, \delta(\varepsilon - \varepsilon_g(\mathbf{k})). \qquad (21b)$$

To proceed, substrate models with specific band structures $\varepsilon = \varepsilon_g(\mathbf{k})$ must be assumed. However, the result given by Eq. (20) is quite general and is not restricted to the p-band metal used for illustrative purposes in Fig. 2.

B. Chemisorption Functions: $\Lambda(\varepsilon), \Delta_a(\varepsilon), \rho_{aa}(\varepsilon)$

For illustrative purposes the level shift function $\Lambda(\varepsilon)$, the level width function $\Delta_a(\varepsilon)$, and the adatom density of states $\rho_{aa}(\varepsilon)$ will be calculated for a simple model, the tight-binding p-band metal shown in Fig. 2. It will be assumed that the nonmagnetic state is appropriate, and the energy $\varepsilon_{a\sigma} = \varepsilon_{a-\sigma}$ given by Eq. (3c) will be treated as a parameter. Note that the value of $\varepsilon_{a\sigma}$ is in part determined by U_e and $\langle n_a \rangle$. To proceed, it is useful to invoke the tight-binding interpolation scheme of Slater and Koster (1954) for obtaining analytic expressions for the energy bands. In this scheme general expressions for the two-center matrix elements are given in terms of parameters determined by the degree of overlap between substrate orbitals on different sites. If the overlap between second nearest neighbors is neglected and if it is assumed that the σ bond between nearest neighbors is much greater than the π bond, then, in the Slater–Koster notation, $\varepsilon_x(\mathbf{k}) \equiv \langle x/x \rangle \simeq p_0 + 2(pp\sigma)_1 \cos(ak_x)$, with p_0 a parameter setting the absolute position on an energy scale and $2(pp\sigma)_1 \equiv p'$ the parameter setting the width of the band. By neglecting second nearest neighbors and retaining only the σ bond between p orbitals, the problem has been reduced to an essentially one-dimensional one. Using this one-dimensional band structure in Eqs. (21a) and (21b), the group orbital resonances, for interstitial adsorbates of p or s character, are given by

$$\rho_p(\varepsilon) = \frac{1}{\pi} \frac{(\varepsilon/p')^2}{(1 - \varepsilon^2/p'^2)^{1/2}} \tag{22a}$$

and

$$\rho_s(\varepsilon) = \frac{1}{\pi}(1 - \varepsilon^2/p'^2)^{1/2}. \tag{22b}$$

Taking $p' = 1$, the nondimensionalized energy band lies in the range $-1 \leqslant \varepsilon \leqslant +1$. With $N_g \beta_g = 1 =$ half the bandwidth, and $\varepsilon_a = 0.25$, the level width functions given by Eqs. (20) and (22a) and (22b) are shown in the top row of Fig. 3. Note that the s adsorbate coupling to states at $k = 0$ and π/a, or at the top and bottom of the band, vanishes due to the destructively interfering relative phases of the host and adsorbate orbital. A similar situation at $k = \frac{1}{2}\pi/a$, or at the middle of the band, occurs for the p adsorbate. The level shift function given by Eq. (4c) is shown in the second row of Fig. 3. In the case of the s adsorbate, the interaction is strongest with the band states in the middle of the band. Consequently the valence level at any $\varepsilon \neq 0$ is repelled from the band center, as can be seen from the $\Lambda_s(\varepsilon)$ curve. On the other hand, the p adsorbate is repelled by the states at the band edges, which results in a piling up of perturbed states near the band center. These notions are graphically illustrated in the third row of Fig. 3, where the adatom density of virtual states from Eqs. (4c), (8), (20), and (22a) and (22b) are shown. In the case of the p adsorbate, ρ_{aa} exhibits a peak at $\varepsilon \equiv \varepsilon_r = 0.125$, which would appear as an extremely narrow resonance, whereas ρ_{aa} for the s adsorbate is spread rather uniformly throughout the whole band. As first noted by Newns (1969a), dramatic differences in ρ_{aa} can occur as a result of small changes in the chemisorption parameters, such as the value of ε_a or β_g. In the present case these parameters are kept constant; still, striking differences in $\rho_{aa}(\varepsilon)$ are present, solely as a result of the orbital symmetries that give rise to sampling of different regions of the Brillouin zone in the V_{ak} mixing. Similar differences should occur in realistic three-dimensional calculations as the bond site, i.e., bridge, four-fold, or head-on, is varied. In any event, one would be hard pressed to characterize these densities of states in terms of a two-parameter Lorentzian, the parameters being the same for both s and p adsorbates. Unfortunately, this increases the difficulty in interpretation of experimentally observed (Gadzuk and Plummer, 1973; Eastman and Cashion, 1971; Waclawski and

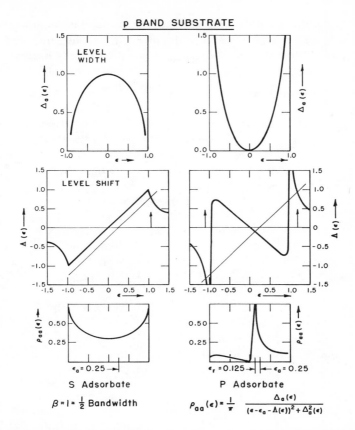

Fig. 3. Various calculated adsorption functions for s and p adsorbates interstitially adsorbed on a tight-binding p-band metal. The band falls in the range $-1 \leqslant \varepsilon \leqslant 1$ (nondimensionalized). The unperturbed adsorbate level is $\varepsilon_a = 0.25$ and the hopping integral $N_g \beta_g = 1$ = half the bandwidth. Top: Level width functions. Middle: Level shift functions. Bottom: Adatom virtual density of states within the band. Note that the intersection of the line $\varepsilon - \varepsilon_a$ with $\Lambda(\varepsilon)$ in the middle curve, for ε outside the band, gives the eigenvalue of the localized states.

Plummer, 1973; Feuerbacher and Fitton, 1973; Hagstrum, 1972) virtual states of chemisorbed atoms if the coupling matrix element $N_g \beta_g$ is comparable with half the bandwidth and if the states are not split off from the band thus forming true localized states.

From Eq. (7), true localized states form when $\varepsilon - \varepsilon_a = \Lambda(\varepsilon)$ outside the band. The energies at which this occurs are marked by arrows on the $\Lambda(\varepsilon)$ curves in Fig. 3. Apparently the spectral weight associated with the state obtained from the adatom level has been distributed between localized states of fractional occupancy below and above the p band and a virtual state distributed within the band. In the case of the p adsorbate, the virtual state displays a resonance half way between the middle of the band and ε_a. Further implications of localized states will be discussed in the next section.

C. Surface Molecule Limit

Considerable simplification in the theory of chemisorption occurs when the adatom–substrate bond is determined mostly by the strong interaction between the adatom orbitals and the localized group orbital formed from nearest neighbor substrate orbitals for states in a tight-binding band, such as the d band of a transition metal. If the interaction is strong enough or the apparent d-band width small enough, then localized states are formed outside the narrow d band centered at ε_d in Figs. 1 and 4. These states correspond to a bonding (ε^-) and antibonding (ε^+) molecular orbital formed from the atomic orbitals of the adsorbate atom and the nearest neighbor substrate atoms. The shift in the molecular levels from ε_a or ε_d is determined by the magnitude of $N_g \beta_g$, and it is this quantity that is most important in determining chemisorption energies.

The surface molecule eigenvalues can be obtained from the Anderson model as follows. As noted by Grimley (1971a, b, 1972, 1973) and Thorpe (1972), the group orbital resonance $\rho_g(\varepsilon)$ given by Eq. (20) is often a strongly peaked function at some energy within the d band, call it ε_d. Replacing $\rho_g(\varepsilon)$ by a delta function,

$$\Delta_g(\varepsilon) \simeq \pi |N_g \beta_g|^2 \delta(\varepsilon - \varepsilon_d),$$

which, when inserted in Eq. (4c), yields

$$\Lambda(\varepsilon) = \frac{|N_g \beta_g|^2}{\varepsilon - \varepsilon_d}.$$

The surface molecule eigenvalues, $\varepsilon = \varepsilon_a + \Lambda(\varepsilon)$ are thus

$$\varepsilon^{\pm} = \frac{\varepsilon_a + \varepsilon_d}{2} \pm \frac{1}{2}((\varepsilon_a - \varepsilon_d)^2 + 4 |N_g \beta_g|^2)^{1/2}, \tag{23}$$

the familiar bonding and antibonding molecular levels shown in Fig. 4. Next the discrete surface molecule state must interact with the remaining indented solid. Since ε^+ and ε^- lie outside the d band, the most likely interaction is a weak residual decay into the s-band continuum as shown in Fig. 4. This converts the discrete surface molecule states into narrow virtual molecular states (Gadzuk, 1973; Grimley and Torrini, 1973). It is hypothesized that this mechanism, in addition to those discussed in Section III.B, can account for the discrepancy between binding energies and level

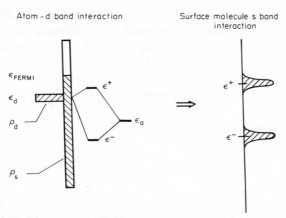

Fig. 4. Surface molecular energy level diagram depicting the formation of localized states on the left due to the interaction of the atomic state at ε_a' with the d-band states at ε_d. The weak interaction of the molecular orbital states ε^- and ε^+ with the s-band continuum converts the discrete states to virtual molecular states, as shown on the right.

widths observed experimentally. The binding energy is supplied in the surface molecule formation, whereas the level width is acquired in the weak residual interaction with the s-band continuum. A chemisorption state in which binding energies are greater than 3 eV and virtual level widths are ~ 1 eV is the type of state consistent with the virtual surface molecule picture. Specific examples of surface molecules will be discussed in Section V.

IV. ELECTRON GAS THEORIES

A. Dielectric Response

Some of the earliest theories of chemisorption were based upon the Born–Haber cycle, which provided a correlation between adatom ionization energies, substrate work function, and heats of adsorption (DeBoer, 1956). As envisioned, an atom far from the substrate is ionized requiring energy V_i. The electron is given to the substrate, thus releasing energy ϕ_e. The resulting ion is brought to the surface, interacting via the attractive classical image potential $V_{im}(z) = -e^2/4z$. Finally the combined system relaxes to its equilibrium configuration, with an energy E_{rel} associated with this stage. By completing the thermodynamical cycle, the atomic chemisorption energy $\equiv E_{initial} - E_{final}$ is given by

$$E_{chem} = \phi_e - V_i + e^2/4z_i + E_{rel}. \tag{24}$$

In actual practice z_i has been taken equal to the "ionic radius" and E_{rel} has been neglected (Rasor and Warner, 1964). Within this model, each adatom has a dipole moment $\mu = 2er_i$ and, from the Helmholtz capacitor equation, the change in work function due to the surface double layer is given by $\Delta\phi = 2\pi\theta\sigma_1\mu$, with θ the fractional coverage and σ_1 the density of adatoms at monolayer coverage. Although crude, this picture illustrates the possible importance of the image force in chemisorption and thus the need for improvement of image force concepts when considering charged particle–surface separations on the level of atomic dimensions. It is interesting to note that the so-called proton model of hydrogenation in metals (Friedel, 1972) is very similar to the Born–Haber model of chemisorption.

One of the fundamental problems with the classical image force is the assumption that the substrate is a perfect conductor and thus the polarization or screening charge induced by the ion is pure surface charge. In fact, the unscreened ion core field penetrates into the solid (or bounded electron gas) on the order of a screening length, and the resulting screening charge is spatially distributed within a layer of this thickness. For equilibrium ion–surface separations, the finite field penetration effects substantially alter both the form of the interaction potential and the resulting dipole moment per adsorbate. Gomer and Swanson (1963) compensated for the finite screening by replacing the image potential with $V_s(z) = e^2/4(z + k_{FT}^{-1})$, where k_{FT}^{-1} is the Fermi–Thomas screening length. A potential of this form has well-behaved limits for both large z going to the classical limit and small positive z going to a finite value. Subsequently, the justification of such a potential has been given (Gomer and Swanson, 1963; Newns, 1969b; Gadzuk, 1970a; Antoniewicz, 1972).

Having recognized the importance of finite polarization effects, several theories of the static dielectric response of a surface to a massive point charge have been formulated (Gadzuk, 1967b, 1969; Gerlach, 1969; Strässler, 1969; Newns, 1970a; Sidyakin, 1970; Beck et al., 1970; Beck and Celli, 1971; Peuckert, 1971; Rudnick, 1972; Heinrichs, 1973). Although these theories differ in some technical points, they are similar in the following ways. The desired quantity is the static screening charge density given in linear response theory by

$$\delta n(\mathbf{r}) = \int d^3r' \, R(\mathbf{r},\mathbf{r}') \, V(\mathbf{r}'), \qquad (25)$$

where $R(\mathbf{r},\mathbf{r}')$ is the response function (simply related to the wave number dependent dielectric function in the case of an infinitely extended electron gas) and $V(\mathbf{r}') = U(\mathbf{r}') + \phi(\mathbf{r}')$, where ϕ is the potential due to the charge density δn and U is the source potential from the ion core. Calculations in most cases are done on a jellium model with an infinitely large surface barrier thus allowing the use of sine wave functions to characterize the motion normal to the surface. The random phase approximation is also invoked. Due to the lack of translational invariance in the surface region, the response function $R(\mathbf{r}, \mathbf{r}') \neq R(\mathbf{r} - \mathbf{r}')$ and the real space problem cannot be Fourier–transformed and easily solved in the way that the impurity in an electron gas problem has been (Langer and Vosko, 1960). The several theories mentioned offer different approximations for handling this problem. In all cases the qualitative conclusions reached are similar. For the ion far removed from the surface, the screening charge appears as a charge sheet on the surface giving rise to the image force. As the ion core moves into the surface region, the finite depth of the screening charge becomes significant; and as the ion penetrates the surface, it becomes embedded in its own screening charge, which ultimately becomes spherically symmetric about the ion core.

Lang and Kohn (1970, 1971, 1973; Lang, 1973) have treated the dielectric screening problem quite differently. Using the Hohenberg–Kohn–Sham (Hohenberg and Kohn, 1964; Kohn and Sham, 1965) theory of the inhomogeneous electron gas adapted to surface phenomena, they have performed self-consistent calculations for the screening charge profiles associated with a uniform charge sheet outside a jellium substrate. In Fig. 5, these profiles are shown along the surface normal. The step edge of the positive jellium background is at $x = 0$, and the jellium extends into the $x < 0$ half space. The center of mass of the screening charge is located at x_0. Three curves are shown for decreasing electron gas densities n, with $n = (\frac{4}{3}\pi a_0^3 r_s^3)^{-1}$ defining the nondimensional parameter r_s. The important point to be made is that as r_s increases, the electron gas density decreases, and the resulting jellium becomes a less ideal conductor. As a consequence, the fields from the source penetrate further into the jellium. Thus the induced screening charge appears further into the substrate as is clearly evidenced in Fig. 5, although the position of the image plane x_0 remains fixed. This is the type of behavior that was qualitatively demonstrated with the previous nonself-consistent dielectric response theories (Gadzuk, 1967, 1969a, 1970a; Antoniewicz, 1972; Gerlach, 1969; Strässler, 1969; Newns, 1970a; Sidyakin, 1970).

Smith et al. (1973) have followed a similar tack in calculating the energy and charge density for the system of semi-infinite jellium plus a proton as a function of proton–jellium separation. A typical screening charge density plot is shown in Fig. 6. The proton is located at $x = 4.5$, $u = 0$ (in units of Bohr radii). Note that the maximum in the screening charge occurs at $x \simeq 3$, and the spread in the screening charge in the transverse directions is much greater than the spread in the x direction. This closely represents a surface charge density. On the other hand, Smith et al. find that by the

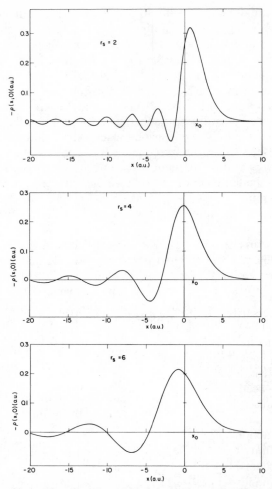

Fig. 5. Profiles of induced surface charge density $\rho(x;0)$ versus x. The edge of the uniform positive-charge background is taken at the origin. x_0, the center of mass of $\rho(x,0)$, is the effective location of the metal surface (Lang and Kohn, 1973).

time the proton sits at $x \simeq 1.5$, the screening-charge is surrounding the proton (much as an atomic orbital associated with a virtual bound state) in a slightly polarized spherical distribution. At this separation, in which the proton is still "outside" the metal, the proton has already caught up with its screening charge, and it becomes somewhat problematic whether to consider the resulting composite as a proton plus screening charge or as an atom or negative ion. Two points should be mentioned concerning the

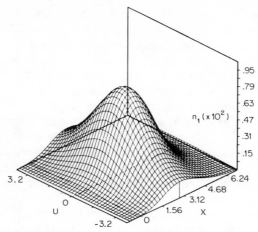

Fig. 6. Screening charge density $n(u, x; x^1)$ for the proton at $u = 0$, $x^1 = 4.5$ (atomic units). The metal extends into the $x < 0$ half space. (From Smith *et al.*)

electron gas screening theories. First, the theories are linear response theories that necessarily require the ratio $\delta n(\mathbf{r})/n_0(\mathbf{r}) \ll 1$, where $n_0(\mathbf{r})$ is the unperturbed local density of background charge. In the surface region $n_0(\mathbf{r})$ rises from zero to its internal density, and in at least part of this region the screening to background charge ratio may be greater than unity. The effects of this breakdown on the significance of the calculated screening charge have not been established.[†] A more serious problem occurs when this formalism is used to calculate the total interaction energy as a function of hydrogen–metal separation. As calculated in the paper by Smith *et al.* (1973), the interaction energy becomes more attractive as the proton is brought from infinity towards the metal, and a minimum in the interaction energy curve occurs at ~ 0.5 Å outside the jellium positive background step edge. For smaller separations a repulsive potential energy dominates, and the total interaction energy curve rises. The problem with this result is that many materials form metal hydrides (Gibb, 1962), where chemisorbed hydrogen is rapidly absorbed into the bulk. In order for this to occur, the total interaction curve must again become attractive at small hydrogen–

[†] Fully self-consistent, exact model calculations for H, Li, and O interacting with a jellium surface have been presented by Lang and Williams (1975) within a local density approximation for exchange and correlation. These calculations confirm the fears expressed in the text that linear response theory in the surface region leads to qualitatively misleading numbers and conclusions. Values for chemisorption energies and dipole moments obtained by Lang and Williams are in much better accord with intuition and with experimental values.

metal separations. Since this does not occur in the jellium theories, some essential interaction mechanism appears to be missing. Whether hydrides form or not depends upon the strength of this additional mechanism (since the only substrate variable in the jellium theory is r_s). Even in cases of no hydride formation, the interaction energy near the surface is likely to be altered significantly. Thus it might be misleading to interpret real chemisorption data on the basis of jellium calculations.

As noted by Applebaum (Volume I), recent theory (Smith *et al.*, 1973; Lang and Kohn, 1970, 1971, 1973) indicates that the effective image plane appears "outside" the metal. Thus the image potential should be of the form $V_{im}(z) = e^2/4(z - z_0)$, whereas the previously mentioned suggestion placed the image plane into the metal and thus $V_{im}(z') = e^2/4(z' + z_0)$. There is no inconsistency in this confusing point, because each theory is referencing the z origin to a different point. The self-consistent theories place $z = 0$ either at the last plane of ion cores or at the jellium step edge. Thus the effective image plane lies outside this point. On the other hand, in a model in which the electron density is a step function at $z = 0$ (Newns, 1969b) or in which an infinite surface potential is assumed at $z = 0$ (Gadzuk, 1970a), the effective image plane is in the negative z region. However, if the step edge of the background charge were located, it too would be in the negative z region, in such a position that the image plane would lie between the step edge and the nodal plane of the sine wave functions. The choice of origin, and not any essentially different physics, dictates the sign in front of z_0 in the image law.

Finally, the role of image potential shifts in adatom energy levels should be mentioned (Bennett and Falicov, 1966; Gadzuk, 1967, 1969; Gadzuk *et al.*, 1971; Muscat and Newns, 1972; Gomer and Swanson, 1963; Remy, 1970). In these papers it has been demonstrated that a hydrogenic atom will have its valence level raised in energy due to the interaction between the valence electron and its image charge and also the image charge of the ion core. From another point of view, Smith *et al.* (1974) have shown that immersing a proton in a surface electron gas converts the bare Coulomb potential into a Yukawa potential. Since the Yukawa potential localizes the bound electron more strongly, the binding energy decreases or, alternatively, the valence level is raised in energy. The image charge shift contains exactly the same physics. The valence electron responds to the total induced screening charge in a repulsive manner that serves to raise the energy level. Although in the Anderson model explicit reference to screening charges is not made, the raising of ε_{a0} to $\varepsilon_{a\sigma}$ occurs for the same basic reason that the combined image potentials raise the valence level. Electrons with $-\sigma$ attempt to "screen" the proton. The electrons with $+\sigma$ see this repulsive "screening" charge and thus their energy increases.

B. Thin Film Model

Up to this point, the models that have been proposed are atomistic and are, strictly speaking, valid only in the extreme low coverage limit where adatom–adatom interactions are negligible. An alternative that has been proposed (Gadzuk, 1970b; Lang, 1971; Newns, 1972) is a bimetallic continuum model in which the adsorbed layer is considered to be a jellium thin film bound by the substrate on one side and by the vacuum on the other. The validity of such a model improves as the thickness of the film increases. Although it might seem inappropriate to envision a partial monolayer film within a continuum model, experimental data, such as surface plasmon energies (Gadzuk, 1970b; MacRae et al., 1969; Thomas and Haas, 1972) and work function changes (Lang, 1971) as a result of adsorbed alkali films, can be understood with this picture.

Following Lang (1971), the main features of the model are displayed in Fig. 7. The substrate, characterized by a positive background charge $\bar{n}_{\text{sub}}$, fills the half space $x < 0$. The adsorbed layer, in this case a full monolayer of adsorbed Na, is replaced by a homogeneous positive slab of thickness $d(\simeq 6a_0$ for the case shown) and density $\bar{n}$. The adsorbate parameters are related to coverage θ through $\bar{n}d = \theta\sigma_1$, with σ_1 the adatom density at monolayer coverage. Variations in θ can be handled by varying either $\bar{n}$ or d, although in practice d is set equal to the inner planar spacing of the bulk adsorbate material and $\bar{n}$ varies as θ does. Using the Hohenberg–Kohn–Sham formalism (Hohenberg and Kohn, 1964; Kohn and Sham,

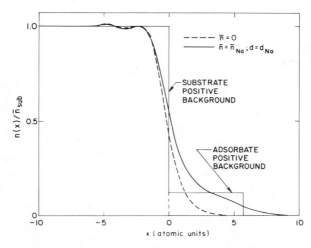

Fig. 7. Self-consistent electron density distributions $n(x)$ for the bare-substrate model ($r_s^{(\text{sub})} = 2$) and for the model of substrate with a full layer of adsorbed Na atoms (Lang, 1971).

1965), the calculated (Lang, 1971) self-consistent charge density for the total system is also shown in Fig. 7. The dashed curve represents the charge density for the clean surface and the solid curve the charge density for the substrate–thin-film–vacuum system. The "screening charge" would be obtained by subtracting the dashed curve from the solid one, and the resultant distributions would resemble those shown in Fig. 5. It is worth noting that when $\bar{n}/\bar{n}_{\text{sub}} < 1$, charge flows from the adsorbate to the substrate, creating an additional dipole layer that lowers the work function of the composite surface. A quite extensive list of experimental work function versus coverage studies is given by Lang (1971) and will not be repeated here. As the continuum model is most valid in the high coverage range, it complements the low coverage atomic level theories. Further implications of this model will be discussed in Section V.B.

V. APPLICATIONS

Up to this point the emphasis has been on general mechanisms of chemisorption rather than on how these mechanisms apply to specific situations. We will now turn to three different sets of systems that have been studied extensively with various experimental techniques and see how the theoretical models discussed here are useful in understanding the data. By no means is this section intended to be a complete survey of the field but is rather a sampling illustrating the constructive relationship between experiment and theory.

A. H on (100) W

One of the most investigated chemisorption systems is dissociatively adsorbed H_2 on the (100) face of W. An excellent summary of the non-spectroscopic experimental data has been given by Madey (1973).

Starting from the Anderson model, there has been theoretical speculation as to the electronic structure of chemisorbed H on transition metal surfaces (Grimley, 1972; Newns, 1969b; Gadzuk, 1973). These developments were in part stimulated by the field emission resonance tunneling studies of Plummer and Bell (1972), who observed a virtual energy level roughly 0.6 eV wide centered at ~ 1 eV beneath the Fermi level. A number of possible surface molecular energy level diagrams consistent with the theory in Sections II and III and also with the field emission results are shown in Fig. 8. A non-magnetic virtual atomic state is shown in Fig. 8a. The virtual level forms from the coalescence of the valence and affinity level and is thus capable of up to

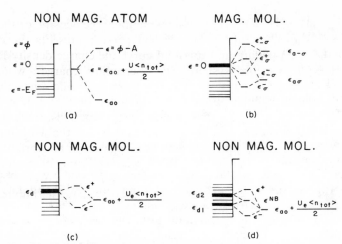

Fig. 8. Possible H–W surface molecule energy levels. (a) Nonmagnetic virtual atomic level. (b) Magnetic surface molecule levels. (c) Nonmagnetic surface molecule levels with a single group orbital resonance. (d) Nonmagnetic surface molecule levels with a double group orbital resonance.

double occupancy. Reasonable values of $\langle n_{tot} \rangle > 1$ place the level in the range expected from the field emission studies, but it is difficult to imagine how the state shown in Fig. 8a is energetically favorable.

Grimley (1972) has proposed the magnetic surface molecule scheme shown in Fig. 8b. His preliminary group orbital calculations for H in the bridge bond site on (100) W indicate that a group orbital resonance, given by Eq. (20), might appear right at the W Fermi level. The self-consistency condition Eq. (9b) together with Eqs. (3a) and (23a) requires

$$\langle n_\sigma \rangle = \left. \frac{\varepsilon^2}{\varepsilon^2 + |N_g \beta_g|^2} \right|_{\varepsilon = (\varepsilon_{a0} + U \langle n_{-\sigma}\rangle/2) \pm \frac{1}{2}((\varepsilon_{a0} + U\langle n_{-\sigma}\rangle)^2 + 4|N_g\beta_g|^2)^{1/2}}.$$

For reasonable choices of $N_g \beta_g$, there are two magnetic solutions to this equation, one for the plus and one for the minus sign in the expression determining ε. This results in two sets of bonding and antibonding molecular levels, one for spin σ and one for spin $-\sigma$, as shown in Fig. 8b. In addition to the spin-down level $\varepsilon^-_{-\sigma}$, which is ~ 1 eV below the Fermi level, a spin-up bonding level $\varepsilon^-_{+\sigma}$ is predicted at $\varepsilon^-_{+\sigma} \sim -6$ eV. However, the reliability of the magnetic HF solution to the Anderson model is not clear.

Model calculations seem to indicate that if a group orbital resonance occurs, it often occurs at the same energy as the maximum in a density of states calculation. The uv photoemission experiments of Plummer and Waclawski (1975) on clean (100) W show a large d-band peak ~ 2 eV below the Fermi level. If the group orbital resonance occurs at this energy and screening effects reduce U to $\sim 5\,\text{eV}$, as discussed in Section II, then the simple energy level diagram shown in Fig. 8c can account for the field emission level and can also provide a satisfactory binding energy. Such an adsorbed H would in fact possess one excess negative charge, as is quite common in volume metal hydrides (Gibb, 1962). It is predicted that this nonmagnetic surface molecule would also have a bonding level at $\varepsilon^- \sim -6\,\text{eV}$.

If the group orbital resonance is also peaked at $\varepsilon \simeq -4.5\,\text{eV}$ where another weaker d-band structure is observed (Plummer and Waclawski, 1975), then the expected surface molecule energy level diagram would be shown in Fig. 8d. In addition to the bonding and antibonding levels at ~ -6 and $-1\,\text{eV}$ respectively, a no-bond level somewhere in the range $-4.5\,\text{eV} \lesssim \varepsilon \lesssim -2\,\text{eV}$ is now apparent. Still, the hydrogen would possess two electrons or one net negative charge.

Recently both Plummer and Waclawski (1975) and also Feuerbacher and Fitton (1973) have studied the electron energy level spectrum of H on (100)W, as a function of coverage, using uv photoemission. Energy distributions with and without H are obtained, and the results are presented as the difference between the two. Presumably any structure in the difference curves is due to photoemission from the virtual levels associated with the H–W bond. Difference curves of Plummer and Waclawski are shown in Fig. 9 with hydrogen exposure (and thus work function changes) the parameter. It is known that the work function change varies linearly with coverage (Madley and Yates, 1970), with $\Delta\phi \simeq 0.75$–$0.8\,\text{eV}$ for a complete monolayer. Concentrating on the low coverage curves, three levels are seen at $\varepsilon = -1.2$, -3.5, and $-5.7\,\text{eV}$. The negative dip at $\varepsilon \simeq -0.4\,\text{eV}$ is due to a decrease in surface state emission (Plummer and Gadzuk, 1970; Waclawski and Plummer, 1972; Feuerbacher and Fitton, 1972) and is of no concern here. Comparing these results with the energy level diagrams in Fig. 8, it is seen that the scheme proposed in Fig. 8d represents the data quite well. Upon modification, the mechanism proposed in Fig. 8c is also consistent with the data. If instead of a delta function group orbital resonance at $\varepsilon = \varepsilon_d$ a more realistic function obtained from Eq. (20) were used, the resulting density of states would then be similar to that shown in the right-hand column of Fig. 4. The 1s H orbital has the same phase relative to the $d_{x^2-y^2}$ metal orbitals with which it would bond while in the bridge site (Estrup and Anderson, 1966) as the p orbital on the p-band substrate in Fig. 4. This is the reason for choosing the right-hand column. In this picture two localized states are split off from the band, and a third maximum within the band is present. The density of localized states would be converted into narrow but symmetric Lorentzian profiles

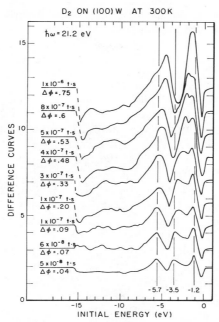

Fig. 9. Photoemission energy distribution difference curves for D_2 on (100) W at 300° K with $\hbar\omega = 21.2\,\text{eV}$. Although small differences in sticking probabilities between D_2 and H_2 exist, the curves are expected to be similar for H_2. The curve labeled $\Delta\phi = 0.75$ corresponds to one monolayer. The energies are referenced to zero at the Fermi level (Plummer and Waclawski, 1975).

due to the residual s-band decay, whereas the d-band virtual state peak is very asymmetric. It is intriguing that the experimentally observed peaks at $\varepsilon = -1.2$ and $-5.7\,\text{eV}$ are symmetric, whereas the middle peak at $\varepsilon = -3.5\,\text{eV}$ displays the same skewed shape as the virtual state in Fig. 4.

In either case, Fig. 8c or d, binding energies $\sim 3\,\text{eV}$ and positive work function changes are predicted, in agreement with experiment (Madey, 1973; Madey and Yates, 1970; Toya, 1972).

The problem of adatom–adatom interactions has also been examined (Einstein and Schrieffer, 1973; Adams, 1974; Schmidt and Gomer, 1966; Sidorski, 1972). Grimley and Torrini (1973) have included the effects of adatoms interacting indirectly through the substrate. They find a mild repulsion between H atoms adsorbed on neighboring sites, which is consistent with the experimental observation of a $c(2 \times 2)$ LEED structure (Estrup and Anderson, 1966) at low coverages. Einstein and Schrieffer (1973) have completed a more extensive theory of indirect interactions between adsorbed H atoms and are able to predict adlayer structure on simple s-band cubium that is consistent with LEED observations on complex d-band transition metals.

B. Alkali Chemisorption

The study of alkali chemisorption on metals has been particularly extensive due to the useful ability of a partial monolayer of adsorbed alkali atoms to reduce the work function of the composite surface to a value lower than that of the alkali solid. The theoretical aspects have been discussed at great length (Gyftopoulos and Levine, 1962a, b; Bennett and Falicov, 1966; Gadzuk, 1967, 1969, 1970b; Gadzuk *et al.*, 1971; Muscat and Newns, 1972; Rasor and Warner, 1964; Gomer and Swanson, 1963; Remy, 1970; Lang, 1971; Wojciechowski, 1973; Ovchinnikov and Tsarev, 1967; Plummer and Rhodin, 1968), and the experimental data has been considered by Lang (1971).[†]

The basic distinguishing property of alkali chemisorption is due to large charge transfer from the alkali to the substrate, thus resulting in an ionic bond. Since the ionization potentials of Cs, K, and Na are 3.89, 4.32, and 5.12 eV respectively, the energy of the valence level, which is shifted upward by image forces and U_e in Eq. (3c), lies above the substrate Fermi level. Consequently the electron drops to the Fermi level, and an alkali ion with net positive charge (although not of integral value) sits on the surface. As treated in Section IV.A, the charged particle polarizes the electron gas of the substrate inducing a screening charge. The resulting dipole configuration reduces the surface barrier for electron emission. Using an Anderson model with electron gas screening effects, the dipole moment per adatom, and thus the initial slope of work function versus coverage curves, has been calculated for the alkalis on various crystal faces of W and Ni (Gadzuk, 1967, 1969). Within the same model, chemisorption energies have also been calculated (Gadzuk *et al.*, 1971). Tolerable agreement with experiment was achieved.

Lang (1971) has utilized the thin film model discussed in Section IV.B in a calculation of work function change versus coverage over the full range of coverage, and a typical result is shown in Fig. 10 for the case of Na on W. This curve demonstrates the common type of result, in which initially ϕ decreases linearly, before adatom–adatom interactions are

---

[†] A definitive experimental and interpretive paper by Andersson and Jostell (1975) has appeared in which Na and K adsorption on Ni(100) has been studied by means of electron energy loss spectroscopy in conjunction with work function measurements and LEED. The interpretation of their data is totally consistent with the theories of alkali chemisorption discussed here.

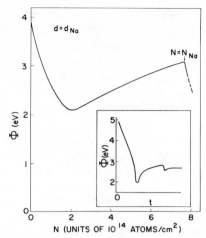

Fig. 10. Computed $\phi(N)$ curve for Na adsorption showing the maximum occurring at the commencement of second-layer formation. The inset shows an experimental curve (Ovchinnikov and Tsarev, 1967) exhibiting such a maximum. The abscissa of this curve is time of exposure t of the W(110) substrate to a flux of Na atoms. The coverage N is presumably a monotonically increasing, though not linear, function of t (Lang, 1971).

important, attains some minimum value less than the work function of the alkali, and then rises. In the theory, second layer formation is not allowed to start until the first layer is completed at $N = N_{Na}$. Once the second layer begins, ϕ drops ultimately to the value characteristic of bulk Na. In experiments, second layer formation generally begins before the first is completed, and thus such structure is washed out. Included in the inset of Fig. 10 is an exception to the general rule, in which the predicted local maximum at full coverage is observed. The experimental results are due to Ovchinnikov and Tsarev (1967). The fact that at $N = 0$ the theoretical bare work function is not equal to the experimental one is of little importance here as present concern is on alkali induced work function changes.

C. 5d Transition Metals on W

An extremely interesting set of experiments were performed by Plummer and Rhodin (1968), in which they studied the binding energies of the 5d transition elements adsorbed on four different crystal faces of W. The general trend they observed was that the binding energy varied linearly with the number of unpaired electrons in the 5d shell, assuming that Hund's rule is valid. In other words the energy increased from Hf, with a surface ground state configuration $5d^2 6s^2$, to a maximum for Re, with a $5d^5 6s^2$

configuration, and then decreased linearly. Variations in the absolute value of the binding energy from one face to another were small, but the binding energies for a given adsorbate almost always decreased on different faces in the order (110), (100), (112), and (111). Plummer and Rhodin conclude "that basically the electron configuration of the adatom determines the magnitude of the binding energy but that the atomic structure of the surface dictates the variation with surface crystallography for a given adatom."

Due to the clean systematics of this data, considerable theoretical interest has been stimulated (Thorpe, 1972; Newns, 1970b, c; Allan and Lenglart, 1970; Grimley and Thorpe, 1971a, b; Cyrot–Lackmann and Ducastelle, 1971). Allan and Lenglart and Cyrot–Lackmann and Ducastelle both constructed theories based on the tight-binding method in which correlation effects on the adatom and substrate atoms were neglected. In both cases the form of the binding energy versus number of unpaired d electrons curve was parabolic with its maximum value for W on W.

Newns (1970) and Grimley and Thorpe (1971a) studied the chemisorption energy trends within the Hartree–Fock approximation on the following Anderson Hamiltonian:

$$H = \sum_\sigma \{\varepsilon_a \sum_m n_{m\sigma} + \sum_k \varepsilon_k n_{k\sigma} + \sum_{km} (V_{mk} c_{m\sigma}^\dagger c_{k\sigma} + \text{H.C.})\}$$

$$+ \tfrac{1}{2}U \sum_{\sigma\sigma'} \sum_{mm'} (1 - \delta_{\sigma\sigma'} \delta_{mm'}) n_{m\sigma} n_{m'\sigma'}, \tag{26}$$

where the creation operators $c_{m\sigma}^\dagger$ refer to the adatom d orbitals of energy ε_a and angular momentum m normal to the surface and all other terms are identical to those in Section II. Under the rather drastic assumption that V_{mk} is identical for all m values, using a "rather arbitrary disposition of parameters" (Grimley and Thorpe, 1971a), and invoking a number of other technical approximations, it is found that ΔE varies with N, the number of d electrons, as

$$\Delta E(N) = \Delta E(5) - c(N-5)^2, \tag{27}$$

where c is determined by both U and V_{mk} in Eq. (26). The variation is also parabolic but about Re, the element with $N = 5$, which is also not in accord with experiment. Grimley and Thorp (1971a) also analyzed Eq. (26) within Hartree–Fock but with different technical approximations than those of Newns. By assuming V_{mk} was independent of m, they obtained binding energy versus N curves that had no simple describable shape. They conclude that to "reproduce the experimental values (within HF Anderson) V_{mk} must be allowed to vary also."

Realizing that the Anderson model totally neglects correlation effects on the substrate atoms, Newns (1970c) and also Grimley and Thorpe (1971b) then formulated the transition metal adsorption problem with the Hubbard model (1963), in which electron–electron correlations are treated on an equal footing for both adsorbate and substrate atoms. Newns considered a two-atom limit of the Hubbard Hamiltonian:

$$H = \sum_{\sigma} \left\{ \varepsilon_a \sum_m n_{m\sigma} + \varepsilon_d \sum_d n_{d\sigma} + \sum_{md} (V_{md} c_{m\sigma}^\dagger c_{d\sigma} + \text{H.C.}) \right\}$$

$$+ \tfrac{1}{2} U_a \sum_{\sigma\sigma'} \sum_{aa'} (1 - \delta_{\sigma\sigma'} \delta_{aa'}) n_{m\sigma} n_{m'\sigma'}$$

$$+ \tfrac{1}{2} U_d \sum_{\sigma\sigma'} \sum_{dd'} (1 - \delta_{\sigma\sigma'} \delta_{dd'}) n_{d\sigma} n_{d'\sigma'}, \tag{28}$$

where the operators $c_{d\sigma}$ refer to the d orbitals on the substrate atom with energy ε_d and U_d is the substrate atom Coulomb repulsion integral. Newns again assumes that V_{md} is a constant independent of m. In the limit in which $U_a = U_d \equiv U$, the eigenvalues of Eq. (28) correspond to Hubbard-like solutions, whereas in the limit $U_a = U$, $U_d = 0$, the eigenvalues of Eq. (28) correspond to Anderson solutions. Calculated binding energies versus N are shown in Fig. 11 for both the Hubbard and Anderson limits. The quantity $U/|V_{md}|$ is treated parametrically, and energies are in units of V_{md}. The experimental results (Plummer and Rhodin, 1968) are shown as a dashed line for $N \leqslant 5$ and as a dot-dash line for $N \geqslant 5$. As is apparent, the "almost triangular peaks similar to the experimental results are obtained for U greater than $\sim 4V_{md}$ in the case of the Hubbard model which is expected to be more realistic physically than that of Anderson in this problem" (Newns, 1970c). In fact if $\Delta E(N = 5)$ is fitted to the observed binding energy of 9.3 eV for Re on (100) W, then the values of U and V_{md}, with $U/V_{md} = 4$, correspond to those expected from band structure conclusions. It is concluded that electron correlations play an important role in this problem.

Finally, a cautionary note should be added concerning the intermediate steps linking raw experimental data to sophisticated theories. In the experiments of Plummer and Rhodin (1968), the technique used for determining binding energies was field desorption. With this method one observes the voltage applied between a field ion microscope tip and a flat electrode that is required to field desorb atoms adsorbed on the tip. From the measured voltage one must infer a value for the electric field at the desorption site. Due to uncertainties in the tip radius and inhomogeneities in the field strength across a crystal facet on the tip, relating voltage to field has built-in problems. However, once a value of the desorption field is decided upon, the binding energy is then obtained indirectly through model dependent theories of field desorption relating field, not voltage,

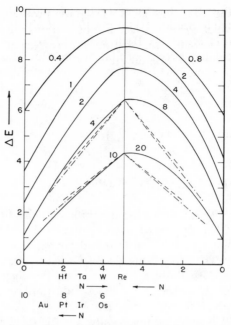

Fig. 11. Binding energies for 5d transition metals on W (in units of V_{md}) versus N for the diatomic Hubbard (left) and Anderson (right) models. Curves are numbered by values of U/V_{md}. Relative experimental binding energies are given by dashed lines as a function of N for $N \leqslant 5$ (upper scale) and by dash-dot lines as a function of N (lower scale) for $N > 5$. The corresponding elements for each N are labeled. Experimental ΔE have been scaled to pass through the maxima of the lower two curves.

to binding energies (Gomer and Swanson, 1963; Muller, 1956; Brandon, 1964). In these theories relatively simple models for chemisorption have been utilized. Thus an inconsistency appears in that detailed theories of chemisorption have been required to explain inferred data in which simple · chemisorption theory was used to relate the observed quantity, voltage, with the inferred quantity, binding energy. It seems reasonable to at least require internal consistency in the process of understanding these intriguing experimental results. This has not yet been achieved. The next meaningful step should be the development of a field desorption theory consistent with the chemisorption theory used to explain the data.

VI. NEW DEVELOPMENTS

Since the completion of this chapter, several papers reporting significant progress in chemisorption theory have appeared. Here we consider the new developments up to July, 1974.

Considerable attention has been directed toward the surface molecule concept, as outlined in Section III (Penn, 1973; Kelly, 1974; Gadzuk, 1974b, Cyrot–Lackmann *et al.*, 1974). Both Penn (1973) and subsequently Kelly (1974) have calculated group orbital resonances (Eq. 20) using more reasonable approximations to the three-dimensional band structure. They find that, in general, although a peak in $\rho_g(\varepsilon)$ might appear within the band, the spectral weight associated with the peak is usually no larger than the spectral weight of the virtual states spread throughout the band. Thus replacing the energy band with a single level at the energy in which $\rho_g(\varepsilon)$ is peaked does not appear to be too reliable for quantitative theory. Gadzuk (1974) and Cyrot–Lackmann *et al.* (1974) have calculated local adatom densities of states for model systems in which no assumptions as to the peaking of $\rho_g(\varepsilon)$ were introduced. As noted in Section III, considerable diversity in $\rho_{aa}(\varepsilon)$ occurs with small changes of system parameters. Doyen and Ertl (1974) have treated the chemisorption of CO on Pt within an Anderson Hamiltonian applied to a small cluster in which the solid state phase relations between orbitals on different substrate centers were neglected. Very nice potential energy surfaces were then calculated.

Several workers have used the surface molecule concept in theories of various spectroscopic studies of surfaces. Einstein (1974) has calculated the change in the local density of states as a function of distance into the substrate, which results when an atom is chemisorbed, and has related this to uv photoemission energy distributions. Gadzuk (1974c, d) and Liebsch (1974) have shown how chemisorption bonding geometry can be determined from the energy and angle resolved uv photoemission current from surface molecule states. The general theory of spectroscopic experiments on surface molecules has been surveyed (Gadzuk, 1974e).

A point of confusion in terminology may have been created in the past. The term "surface molecule" has sometimes been used to denote the coupled atom–metal states formed in a model in which the d band is replaced by a single energy level, or, in other words, $\rho_g(\varepsilon)$ is taken to be a delta function. This is a more restrictive definition of "surface molecule" than the one currently being used by many workers. The alternative, and hopefully more accepted in the future, definition refers to the localized states that are split off from the d band of the substrate. It is assumed that the adatom bonds to some directed orbitals that are not eigenfunctions but that can be expressed as a coherent superposition of substrate eigenfunctions. This is essentially the scheme outline in Section III. The chemisorption problem is then solved exactly (within the limitations of the imposed model), and localized states may split off from the d-band continuum. It is these states, which are molecular orbital states formed from linear combinations of atomic orbitals on the adsorbate plus a few substrate neighbor atoms, that are now being called surface molecule states. In general, if the adatom valence level energy, Eq. (3c), falls outside the d-band energy range, then such a surface molecule state will be formed.

Anders *et al.* (1973) have carried out a molecular orbital calculation for H adsorbed on (100) W. Energy levels and total energies were calculated for various bond geometries. The energies for the head-on, bridge, and four-fold sites were sufficiently similar that no definitive conclusion concerning the bond geometry could be made.

Deuss and van der Avoird (1973) performed MO cluster calculations on the system of H_2 on various faces of Ni. Potential energy surfaces and reaction coordinates for dissociative chemisorption were obtained. Due to the oversimplified bonding scheme proposed, the relevance of these calculations to real systems is uncertain. However, this beautiful piece of work certainly indicates the direction theoretical work in chemisorption should take in order to fulfill the promise of "understanding" catalysis.

Several papers dealing with more formal aspects of the Anderson–Newns approach have appeared. Madhukar (1973), Brenig and Schönhammer (1974), and Bagchi and Cohen (1974) have developed models in which Coulomb correlations on both the substrate and adsorbate atoms and nonorthogonality between the adatom and substrate wave functions have been included. Within their chosen models, the importance of the relative magnitudes of the intra-atomic Coulomb integral U, the hopping matrix element V, the d-band width W, and the overlap integral S are assessed, and it is shown how bonding goes continuously from Heitler–London to molecular orbital as these parameters vary.

Lastly, the activity in x-α scattered wave calculations should be mentioned. A number of research groups are now working on such calculations (Johnson and Messmer, 1974; Rosch and Rhodin, 1974). The method provides a well-defined prescription for obtaining a numerical result. At this stage, however, the technical problems associated with calculating molecular structure of essentially planar (or very asymmetrical) geometries in terms of spherically symmetric models raise some questions concerning the accuracy of the results.

As can be seen, the field of chemisorption theory is quite active and vital. It is anticipated that many of the problems raised and partially answered in this chapter will receive increasing attention in the next few years, and we soon may actually be in a position to answer the question, "Why does a catalyst catalyze?"

ACKNOWLEDGMENTS

The author has benefitted from contact with many co-workers and colleagues, in particular the long joint effort with Ward Plummer and other NBS colleagues, Ted Madey, David Penn, Bernie Waclawski, and John Yates. Special thanks also to Tom Grimley, Norton Lang, Dennis Newns, Thor Rhodin, Bob Schrieffer, and John Smith for various favors and help.

REFERENCES

ADAMS, D. L. (1974). *Surface Sci.* **42**, 12.
ALLAN, G., and LENGLART, P. (1970). *J. Phys. Suppl. Fasc. 4*, **31**, C193.
ANDERS, L. W., HANSEN, R. S., and BARTELL, L. S. (1973). *J. Chem. Phys.* **59**, 5277.
ANDERSON, P. W. (1961). *Phys. Rev.* **124**, 41.
ANDERSSON, S., and JOSTELL, U. (1975). *Surface Sci.* **46**, 625.
ANTONIEWICZ, P. R. (1972). *J. Chem. Phys.* **56**, 1711.
BAGCHI, A., and COHEN, M. H. (1974). *Phys. Rev. B* **9**, 4103.
BECK, D. E., and CELLI, V. (1971). *Phys. Rev. B* **2**, 2955.
BECK, D. E., CELLI, V., LO VECCHIO, G., and MAGNATERRA, A. (1970). *Il Nuovo Cimento B* **68**, 230.
BENNETT, A. J., and FALICOV, L. M. (1966). *Phys. Rev.* **151**, 512.
BETHE, H. A., and SALPETER, E. E. (1957). "Quantum Mechanics of One- and Two-Electron Atoms." Academic Press, New York.
BRANDON, D. G. (1964). *Surface Sci.* **3**, 1.
BRENIG, W., and SCHÖNHAMMER, K. (1974). *Z. Phys.* **267**, 201.
CLOGSTON, A. M. (1962). *Phys. Rev.* **125**, 439.
CYROT–LACKMANN, F., and DUCASTELLE, F. (1971). *Phys. Rev. B* **4**, 2406.
CYROT–LACKMANN, F., DESJONQUERES, M. C., and GASPARD, J. P. (1974). *J. Phys. C* **7**, 925.
DEBOER, J. H. (1956). *Adv. Catal.* **8**, 17 .
DEUSS, H., and VAN DER AVOIRD, A. (1973). *Phys. Rev. B* **8**, 2441.
DOYEN, G. and ERTL, G. (1974). *Surface Sci.* **43**, 197.
EASTMAN, D. E., and CASHION, J. K. (1971). *Phys. Rev. Letters* **27**, 1520.
EDWARDS, D. M., and NEWNS, D. M. (1967). *Physics Lett.* **24A**, 236.
EINSTEIN, R. L., and SCHRIEFFER, J. R. (1973). *Phys. Rev.* **B7**, 3629.
EINSTEIN, T. E. (1974). *Surface Sci.* **45**, 713.
ESTRUP, P. J. and ANDERSON, J. (1966). *J. Chem. Phys.* **45**, 2254.
FEUERBACHER, B., and FITTON, B. (1972). *Phys. Rev. Lett.* **29**, 786.
FEUERBACHER, B. and FITTON, B. (1973). *Phys. Rev. B* **8**, 4890.
FLAIM, T. A., and OWNBY, P. D. (1972). *Surface Sci.* **32**, 519.
FRIEDEL, J. (1972). *Buns.-Gesellschaft Phys. Chem.* **76**, 828.
GADZUK, J. W. (1967a). *Surface Sci.* **6**, 133, 159.
GADZUK, J. W. (1967b). *Solid State Commun.* **5**, 743.
GADZUK, J. W. (1969a). *J. Phys. Chem. Solids* **30**, 2307.
GADZUK, J. W. (1969b). *In* "The Structure and Chemistry of Solid Surfaces" (G. A. Somorjai, ed.). Wiley, New York.
GADZUK, J. W. (1970a). *Surface Sci.* **23**, 58.
GADZUK, J. W. (1970b). *Phys. Rev.* **B1**, 1267.
GADZUK, J. W. (1973). *Phys. Electron. Conf., 33rd Berkeley, California,* March.
GADZUK, J. W. (1974a). *J. Vac. Sci. Technol.* **11**, 275.

GADZUK, J. W. (1974b). *Surface Sci.* **43**, 44.
GADZUK, J. W. (1974c). *Solid State Commun.* **15**, 1011.
GADZUK, J. W. (1974d). *Phys. Rev. B* **10**, 5030.
GADZUK, J. W. (1974e). *Jap. J. Appl. Phys. Suppl.* **2**, 851.
GADZUK, J. W., and PLUMMER, E. W. (1973). *Rev. Mod. Phys.* **45**, 487.
GADZUK, J. W., HARTMAN, J. K., and RHODIN, T. N. (1971). *Phys. Rev.* **B4**, 241.
GERLACH, E. (1969). *In* "Molecular Processes on Solid Surfaces" (E. Drauglis, R. D. Gretz, and R. I. Jaffee, eds.). McGraw-Hill, New York.
GIBB, T. R. P., Jr. (1962). *Progr. Inorganic Chem.* **3**, 315.
GOMER, R., and SWANSON, L. W. (1963). *J. Chem. Phys.* **38**, 1613.
GOODENOUGH, J. B. (1963). "Magnetism and the Chemical Bond." Wiley (Interscience), New York.
GRIMLEY, T. B. (1971a). *J. Vac. Sci. Technol.* **8**, 31.
GRIMLEY, T. B. (1971b). *Buns.-Gesellsch. Phys. Chem.* **75**, 1003.
GRIMLEY, T. B. (1972). *In* "Adsorption–Desorption Phenomena" (F. Ricca, ed.). Academic Press, New York.
GRIMLEY, T. B. (1973). *In* "The Structure and Properties of Metal Surfaces" S. Shinodaira, M. Maeda, G. Okamoto, M. Onchi, and Y. Tamari eds. Maruzen, Tokyo.
GRIMLEY, T. B., and THORPE, B. J. (1971a). *J. Phys. F* **1**, L4.
GRIMLEY, T. B., and THORPE, B. J. (1971b). *Phys. Lett.* **37A**, 459.
GRIMLEY, T. B., and TORRINI, M. (1973). *J. Phys. C: Solid State Phys.* **6**, 868.
GYFTOPOULOS, E. P., and LEVINE, J. (1962a). *J. Appl. Phys.* **33**, 67.
GYFTOPOULOS, E. P., and LEVINE, J. (1962b). *Surface Sci.* **1**, 171, 225, 349.
HAGSTRUM, H. D. (1972).*Science* **178**, 275.
HEEGER, A. J. (1969). *Solid State Phys.* **23**, 284.
HEINE, V. (1972). *Proc. Roy. Soc.* (London) **A331**, 307.
HEINRICHS, J. (1973). *Phys. Rev. B* **8**, 1346.
HERRING, C. (1966). Exchange interactions among itinerant electrons. *In* "Magnetism" (G. T. Rado and H. Suhl, eds.), Vol. IV. Academic Press, New York.
HOHENBERG, P., and KOHN, W. (1964). *Phys. Rev.* **136**, B864.
HORIUTI, J., and TOYA, T. (1969). *Solid State Surface Sci.* **1**, 1.
HUBBARD, J. (1963). *Proc. Roy. Soc.* (London) **A276**, 238.
JOHNSON, K. H., and MESSMER, R. P. (1974). *J. Vac. Sci. Technol.* **11**, 236.
KELLY, M. J. (1974). *J. Phys. C* **7**, L157.
KJÖLLERSTRÖM, B., SCALAPINO, D. J., and SCHRIEFFER, J. R. (1966). *Phys. Rev.* **148**, 665.
KNOR, Z. (1971). *J. Vac. Sci. Technol.* **8**, 57.
KOHN, W., and SHAM, L. J. (1965). *Phys. Rev.* **140**, A1133.
KOSTER, G. F. and SLATER, J. C. (1954). *Phys. Rev.* **96**, 1167, 1208.
LANG, N. D. (1971). *Phys. Rev.* **B4**, 4234.
LANG, N. D. (1973). *Solid State Phys.* **28**, 225.
LANG, N. D., and KOHN, W. (1970). *Phys. Rev. B* **1**, 4555.
LANG, N. D., and KOHN, W. (1971). *Phys. Rev. B* **3**, 1215.
LANG, N. D., and KOHN, W. (1973). *Phys. Rev. B* **7**, 3541.
LANG, N. D., and WILLIAMS, A. R. (1975). *Phys. Rev. Lett.* **34**, 531.
LANGER, J. and VOSKO, S. J. (1960). *J. Phys. Chem. Solids* **12**, 196.
LIEBSCH, A. (1974). *Phys. Rev. Lett.* **32**, 1203.
MACRAE, A. U., MULLER, K., LANDER, J. J., MORRISON, J., and PHILLIPS, J. C. (1969). *Phys. Rev. Lett.* **22**, 1048.
MADEY, T. E. (1973). *Surface Sci.* **36**, 281.

MADEY, T. E., and YATES, J. T. Jr. (1970). *In* "Structure et Propriétés des Surfaces des Solides," No. 187, p. 155. Ed. Centre Nat. Rech. Sci., Paris.
MADHUKAR, A. (1973). *Phys. Rev.* **8**, 4458.
MESSMER, R. P., TUCKER, C. W., Jr., and JOHNSON, K. H. (1974). *Surface Sci.* **42**, 341.
MULLER, E. W. (1956). *Phys. Rev.* **102**, 618.
MUSCAT, J. P., and NEWNS, D. M. (1972). *Solid State Commun.* **11**, 737.
NEWNS, D. M. (1969a). *Phys. Rev.* **178**, 1123.
NEWNS, D. M. (1969b). *J. Chem. Phys.* **50**, 4572.
NEWNS, D. M. (1970a). *Phys. Rev. B* **1**, 3304.
NEWNS, D. M. (1970b). *Phys. Lett.* **33A**, 43.
NEWNS, D. M. (1970c). *Phys. Rev. Lett.* **25**, 1575.
NEWNS, D. M. (1972). *Phys. Lett.* **38A**, 341.
OVCHINNIKOV, A. P., and Tsarev, B. M. (1967). *Fiz. Tverd. Tela* **9**, 1927 [*English transl.:* (1968). *Sov. Phys.-Solid State* **9**, 1519)].
PAULING, L. (1960). "The Nature of the Chemical Bond." Cornell Univ. Press, Ithaca, New York.
PAULSON, R. H., and SCHRIEFFER, J. R. (1975). *Surface Sci.* **48**, 329.
PENN, D. R. (1973). *Surface Sci.* **39**, 333.
PEUCKERT, V. (1971). *Z. Phys.* **241**, 191.
PHILLIPS, J. C. (1969). "Covalent Bonding in Crystals, Molecules, and Polymers." Univ. of Chicago Press, Chicago, Illinois.
PLUMMER, E. W., and BELL, A. E. (1972). *J. Vac. Sci. Technol.* **9**, 583.
PLUMMER, E. W., and GADZUK, J. W. (1970). *Phys. Rev. Lett.* **25**, 1493.
PLUMMER, E. W., and RHODIN, T. N. (1968). *J. Chem. Phys.* **49**, 3479.
PLUMMER, E. W., and WACLAWSKI, B. J. (1975). *Surface Sci.* (in press).
RASOR, N. S., and WARNER, C. III (1964). *J. Appl. Phys.* **35**, 2589.
REMY, M. (1970). *J. Chem. Phys.* **53**, 2487.
RÖSCH, N., and RHODIN, T. N. (1974). *Phys. Rev. Lett.* **32**, 1189.
RUDNICK, J. (1972). *Phys. Rev. B* **5**, 2863.
SCHRIEFFER, J. R. (1972). *J. Vac. Sci. Technol.* **9**, 561.
SCHRIEFFER, J. R., and GOMER, R. (1971). *Surface Sci.* **25**, 315.
SCHRIEFFER, J. R., and MATTIS, D. C. (1965). *Phys. Rev.* **140**, A1412.
SCHMIDT, L. D., and GOMER, R. (1966). *J. Chem. Phys.* **45**, 1605.
SIDORSKI, Z. (1972). *Acta Phys. Pol.* **A42**, 437.
SIDYAKIN, A. V. (1970). *Zh. Eksp. Teor. Fiz.* **58**, 573 [*English transl.:* (1970). *Sov. Phys.—JETP* **31**, 308].
SLATER, J. C., and JOHNSON, K. H. (1972). *Phys. Rev.* **B5**, 844.
SLATER, J. C., and KOSTER, G. K. (1954). *Phys. Rev.* **94**, 1498.
SMITH, J. R., YING, S. C., and KOHN, W. (1973). *Phys. Rev. Lett.* **30**, 610.
SMITH, J. R., YING, S. C., and KOHN, W. (1974). *Solid State Commun.* **15**, 1491.
STRÄSSLER, S. (1969). *Phys. Kon. Mater.* **10**, 219.
TAMM, P. W., and SCHMIDT, L. D. (1969). *J. Chem. Phys.* **51**, 5352.
THOMAS, S. and HAAS, T. W. (1972). *Solid State Commun.* **11**, 193.
THORPE, B. J. (1972). *Surface Sci.* **33**, 306.
TOYA, T. (1972). *J. Vac. Sci. Technol.* **9**, 890.
WACLAWSKI, B. J., and PLUMMER, E. W. (1972). *Phys. Rev. Lett.* **29**, 783.
WACLAWSKI, B. J., and PLUMMER, E. W. (1973). *J. Vac. Sci. Technol.* **10**, 292.
WEINBERG, W. H., and MERRILL, R. P. (1972). *Surface Sci.* **33**, 493.
WOJCIECHOWSKI, K. F. (1971). *Progr. Surface Sci.* **1**, 65.
WOJCIECHOWSKI, K. F. (1973). *Surface Sci.* **36**, 689.

8

Chemical Analysis of Surfaces

ROBERT L. PARK[†]

SANDIA LABORATORIES
ALBUQUERQUE, NEW MEXICO

I. INTRODUCTION

If one of the surfaces exposed by cleaving a sheet of mica is sprayed immediately with distilled water from an atomizer, the water seems to disappear as it uniformly wets the surface. If several seconds are allowed to pass before the other surface is sprayed, the water will now bead up in distinctly visible droplets. In those few seconds the second surface has been contaminated by the condensation of organic vapors from the laboratory air. Even at a pressure of 10^{-10} Torr, a complete monolayer of contamination can condense on a surface in ~ 2 hours. This, coupled with the tendency of bulk impurities to segregate at boundaries, obliges the surface scientist to

[†] Present address: Department of Physics, University of Maryland, College Park, Maryland.

establish the cleanliness of his surface in advance of every experiment. The ability to determine surface composition with some measure of confidence, however, allows one to do more than simply monitor contamination. Little is presently understood, for example, of the factors that determine the surface composition of alloys in a reactive environment, despite the importance of this problem to lubrication, catalysis, corrosion, etc.

The field of surface chemical analysis can be properly said to have begun when Harris (1968) demonstrated that Auger emission peaks could be sensitively detected in the secondary emission spectrum of an electron-bombarded surface by the simple expedient of plotting its energy derivative. Electron-excited Auger spectroscopy was eagerly greeted by surface scientists who had been haunted by uncertainty over the cleanliness of their samples. Since that time a wide assortment of methods for the elemental analysis of solid surfaces has become available.

All of these methods depend on a determination of either the mass or the charge of the atomic nucleus. These basic properties are deduced by a seemingly endless concatenation of stratagems, each of which views the surface from a somewhat different perspective. The task of the surface physicist is to reconcile these different viewpoints. It is a task that has hardly been approached. The field of surface chemical analysis is, after all, of quite recent origin, and although it would be a profound mistake to suppose that no new spectroscopies will be discovered, there is much progress to be made in better utilization of those we have. There is more information in our spectra than we presently know how to utilize and few important questions that can be answered without some ambiguity by any single technique.

With this in mind it will be the purpose of this chapter to contrast the principal techniques for surface chemical analysis in terms of the intrinsic differences in the kind of information they provide.

II. NUCLEAR CHARGE

The atomic number Z of a surface atom can be inferred from the energy required to create a vacancy in an inner electron shell or from the energies of the electronic transitions involved in the decay of the vacancy. This decay may be by the emission of characteristic x rays or by the emission of Auger electrons. Although the inner shells of the atoms do not directly participate in chemical bonds, both the excitation and recombination transitions may involve states within a few eV of the Fermi energy. The line shapes associated with these transitions thus reflect the chemical environment of the atoms. The localized core electron, however, overlaps only part of the valence and conduction band density of states. By using the core level as a window, a local density of states associated with a single atomic species is viewed. It is an inevitable consequence of the uncertainty principle that this local density of states will be smeared out relative to that which we would measure by directly probing the valence or conduction bands. The core levels themselves may be measurably shifted relative to the Fermi energy by redistributions of the valence electrons. These chemical shifts are small compared to the binding energies of the core levels and do not seriously interfere with elemental identification. They can, however, provide an important clue to the chemical state of atoms.

A. Excitation Techniques

In principle, any incident particle of energy greater than the binding energy of a core electron can be used to create the initial vacancy. Photons and electrons are the most convenient and least destructive ionizing particles, but protons and heavier ions have been the object of recent interest and, as we shall see, provide still another view of the surface. The reorganization of the excited atom is generally independent of the mode of excitation, since the decay time is long compared to the collision time. The lifetime of the excited state is, however, sufficiently finite to lead to a significant broadening of the level. Thus the width of a core level will not be less than $\sim 0.2\,\text{eV}$. Techniques that study the excitation of the core levels, as opposed to those that examine the decay of the excited state, have been reviewed recently by Park and Houston (1973a).

1. X-RAY PHOTOELECTRON SPECTROSCOPY (XPS)

An accurate knowledge of inner electron levels is fundamental to studies of chemical composition and bonding. The first extensive table of core electron binding energies, published by M. Siegbahn (1931), relied on x-ray absorption measurements to establish a reference level for each element. For $Z > 51$, Siegbahn used the L_3 edge as a reference because of its intrinsic sharpness. For lower Z elements he used the K edge. The energies of the remaining levels were calculated from x-ray emission wavelengths. Siegbahn's table was expanded and revised by Hill et al. (1952) using more accurate values of the physical constants. The principal uncertainty in determining binding energies from the x-ray absorption spectrum results from structure introduced by variations in the density of conduction band states. This is avoided in x-ray photoelectron spectroscopy by exciting the core electrons to states far above the Fermi energy, where the transition density is a slowly varying function of energy. Recent tabulations of binding energies by Bearden and Burr (1967) and K. Siegbahn et al. (1967) have therefore relied on x-ray photoelectron measurements, rather than absorption edges, to establish the absolute scale wherever available. It is not possible, however, to assert unequivocally that this has resulted in more accurate binding energies.

The x-ray photoelectron technique is illustrated by the energy level diagram of Fig. 1. X-ray photons γ of known wavelength are allowed to impinge on the sample surface. If the photon energy $h\nu$ exceeds the binding energy E_B of a core electron, it may be adsorbed by the electron, exciting it into an available state above the Fermi energy E_F. The binding energy of an electron in a solid is by convention taken to be the energy required to excite the electron just to the Fermi energy, in contrast to the free-atom definition, which requires that the electron be removed to infinity. The conventions clearly differ by the work function. If the energy of the ejected core electron exceeds the work function of the sample, $e\varphi_{samp}$, it may be emitted into the vacuum. The XPS experiment consists of measuring the kinetic energies of these photoelectrons.

If we assume, for simplicity, that the surface of the electron spectrometer has a uniform work function, $e\varphi_{spec}$, then the measured kinetic energy E_K of an emitted core electron is given by

$$E_K = h\nu - (E_B + e\varphi_{spec}).\qquad(1)$$

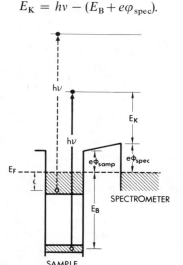

Fig. 1. Energy level diagram of the x-ray photoelectron experiment. A photon of energy $h\nu$ may be adsorbed by an electron in a core level (solid arrow). The measured kinetic energy of the ejected core electron is given by $E_K = h\nu - E_B - e\varphi_{spec}$, where E_B is the binding energy of the core electron relative to the Fermi energy E_F and $e\varphi_{spec}$ is the work function of the electron energy analyzer. The incident photon may instead be adsorbed by a valence electron (dashed arrow), in which case the line shape reflects the density of filled (valence band) states. For a metal, the maximum kinetic energy corresponds to electrons ejected from the Fermi level.

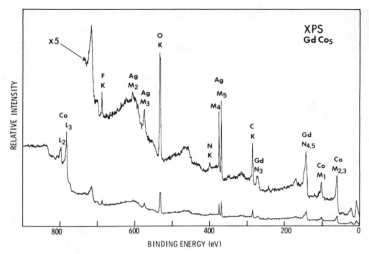

Fig. 2. X-ray photoelectron spectrum of a GdCo$_5$ sample taken by L. H. Scharpen of the Hewlett-Packard Company using the Hewlett-Packard 5950A ESCA spectrometer. The surface was lightly sputtered with argon ions before the spectrum was taken, but substantial amounts of oxygen and carbon remain on the surface as well as traces of nitrogen and fluorine. The origin of the silver contaminant is not known. The broad poorly defined features are probably Auger spectra, but they are difficult to identify. On the low kinetic energy side of each photoelectron peak the background rises as a result of inelastic scattering of escaping photoelectrons. The spectrum was taken in about 3 min.

Although $e\varphi_{\text{spec}}$ is not, in general, precisely known, it should remain constant and relative binding energies can therefore be accurately determined. This assumes, of course, that the ejected core electrons undergo no inelastic collisions on their way out of the sample.

For most ejected core electrons this is clearly not the case. Electron–electron interactions result in strong inelastic–collision–induced attenuation of the intensity of the photoelectrons. It is, in fact, this inelastic collision damping that gives the electron spectroscopic techniques their surface sensitivity, as Duke and Tucker (1969) pointed out for low-energy electron diffraction. The effects of inelastic collisions are evident in the x-ray photoelectron spectrum shown in Fig. 2. On the low kinetic energy side of the photoelectron peak (high binding energy), the background is seen to rise as a result of photoelectrons that have undergone inelastic collisions before escaping from the sample. Palmberg and Rhodin (1968) fixed the mean escape depth of 362 eV Auger electrons in Ag at ∼8 Å. Until 1972, however, the most frequently quoted value for the escape depth of x-ray photoelectrons was 100 Å. This value was determined for multilayers of iodostearic acid (Siegbahn *et al.*, 1967), but unfortunately it was taken as representative of escape depths for all manner of solids.

This had two untoward consequences: Too little attention was given to the cleanliness of surfaces in XPS measurements, and an equivalence was assumed between core electron binding energies measured by XPS and those determined from x-ray adsorption edges. Thus few XPS measurements were taken under the ultraclean conditions required for surface studies, and the effect of reduced coordination on core electron binding energies of surface atoms was ignored. This misconception was expunged by direct measurements of the attenuation of x-ray photoelectrons by graphite (Steinhardt et al., 1972), gold (Klasson et al., 1972), and mercury (Brundle and Roberts, 1973). The attenuation lengths for electrons of 1.4 keV energy are 10–20 Å for these materials, in reasonable agreement with the results of other techniques.

Appreciation of the surface sensitivity of XPS has led to the adoption of standard surface cleaning techniques (Hüfner et al., 1972) and the investigation of XPS as a technique for the study of submonolayer quantities of adsorbates (Brundle and Roberts, 1972; Shön and Lundin, 1973; Madey et al., 1973). The angular dependence of the sampling depth has been examined by Fraser et al. (1973).

Although the application of XPS to surface studies is a recent development, the technique itself is not. It was used as early as 1914 by H. R. Robinson and his colleagues, who obtained spectra from a great variety of materials with a magnetic spectrometer (see, for example, Robinson and Young, 1930). The technique was resurrected and made more sophisticated by K. Siegbahn and his colleagues at the University of Uppsala (Siegbahn et al., 1967), who constructed an iron-free high-resolution spectrometer.

The x-ray source is usually unmonochromatized K_α radiation of Mg(1.25 keV) or A1 (1.49 keV). The natural width of the unresolved $K_{\alpha 1,2}$ doublet is about 1 eV, which is the limiting factor in the resolution of most XPS systems, although in practice the resolution is frequently degraded deliberately to improve the signal–to–noise ratio. The spectra are complicated by the $K_{\alpha 3,4}$ satellites, but these can be eliminated and the achievable resolution improved by the use of monochromatized x-rays (Siegbahn et al., 1972). The GdCo$_5$ spectrum shown in Fig. 2 was obtained by L. H. Scharpen of the Hewlett-Packard Company with a dispersion-compensated electrostatic analyzer and monochromatized x-ray source. The practical resolution that can be achieved with a spectrometer of this sort is about 0.25 eV, which is comparable to the widths of the sharpest core levels.

The width of a core level is a consequence of the finite lifetime τ of the core hole that must be created to detect the level. From the uncertainty principle the width is h/τ. The initial core level must therefore be described by a state distribution function $N_i(E)$, which has the form of a Lorentzian. A full width at half maximum of 1 eV corresponds to a lifetime of 6.6×10^{-16}

sec (Parratt, 1959). The measured line shape may be distorted by dynamic screening of the core hole by the valence electrons (Doniach and Šunjic, 1970).

The structure observed at small binding energies (maximum kinetic energy) corresponds to electrons excited directly from the valence band. This has the advantage over ultraviolet photoelectron studies of the valence band that electrons are excited to states far above the Fermi energy, where variations in state density are negligible. The transition density $T(E)$ should therefore accurately reflect the density of valence band states $N_v(E)$, i.e.,

$$T(E) = \int_0^E N_v(E')\,\delta(E - hv - E')\,dE' = N_v(E - hv). \tag{2}$$

For the purpose of discussion we have assumed a perfectly monochromatic x-ray source and an ideal spectrometer. A number of studies of the density of states have been undertaken for pure metals (Fadley and Shirley, 1968; Baer et al., 1970). The direct excitation of electrons from the valence band of alloys or compounds can be misleading, however, since such fundamental questions as the extent to which the constituents retain their own electronic states cannot be answered unambiguously.

A much more local view of the density of states that can address such questions is provided by the Auger spectrum that results from the decay of the core holes left behind by the photoelectrons. The core levels here serve as windows through which the local density of states associated with a given component can be viewed. X-ray-excited Auger spectroscopy has been largely ignored by those studying surfaces with XPS, but the advantages of having both the excitation and deexcitation spectrum in a single technique will certainly be exploited (Wagner and Biloen, 1973).

2. IONIZATION SPECTROSCOPY (IS)

It is instructive to consider what the problems would be in attempting to obtain the excitation spectrum of the core levels with incident monoenergetic electrons rather than photons. The motivation for such an attempt is the relative simplicity of electron sources. With a simple thermionic emitter and a voltage supply, we have a source of ionizing radiation as monochromatic as any x-ray source and much more intense. Nor is the energy of incident electrons confined to a few characteristic values as are x-ray line sources.

The first problem that must be faced is that the incident electrons, unlike photons, have no hestitation in transferring only a part of their energy to a core electron. This is shown schematically in Fig. 3. V is the potential applied between an emitter (cathode) and the sample. The incident electron, rather than giving up all its energy to the core electron, may be captured in some

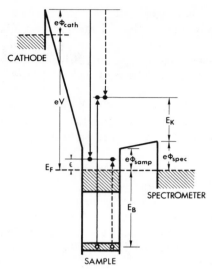

Fig. 3. Energy level diagram of the core excitation characteristic loss experiment (ionization spectroscopy). The energy of an incident electron relative to the Fermi energy is $eV + e\varphi_{\text{cath}}$. The maximum kinetic energy of the ejected core electron is $E_{\text{K}_{\max}} = eV + e\varphi_{\text{cath}} - E_{\text{B}} - e\varphi_{\text{spec}}$ and corresponds to the case where $\varepsilon = 0$. The situation indicated by the dashed arrows is physically indistinguishable from that represented by solid arrows. The shape of the excitation edge reflects the density of unfilled (conduction band) states.

state ε above the Fermi energy of the sample. The kinetic energy of the ejected core electron is

$$E_{\text{K}} = (eV + e\varphi_{\text{cath}}) - (E_{\text{B}} + e\varphi_{\text{spec}}) - \varepsilon. \tag{3}$$

Thus the kinetic energy is not uniquely determined in the electron bombardment case. The maximum kinetic energy is uniquely determined, however, and corresponds to the case where ε is zero, i.e.,

$$E_{\text{K}_{\max}} = (eV + e\varphi_{\text{cath}}) - (E_{\text{B}} + e\varphi_{\text{spec}}). \tag{4}$$

This is the same as Eq. (1), except $h\nu$ has been replaced by $eV + e\varphi_{\text{cath}}$, but it now defines an edge rather than a peak. It should be noted that the situation shown in Fig. 3 by solid arrows is physically indistinguishable from the case in which the core electron is excited to ε, and the incident electron is captured by a state $E_{\text{K}} + e\varphi_{\text{spec}}$ above the Fermi energy (dashed arrows). The probability of finding one or the other of these electrons at a given energy thus depends on the sum of the scattering amplitudes rather than the intensities of these paths.

In a simple one-electron picture, assuming constant oscillator strengths, the probability that the excited core electron will be found at an energy E_{K}

depends on the product of the density of states available at that energy and the density of states at the energy ε defined by Eq. (3). Above perhaps a hundred electron volts, the density of states is so slowly varying that, as in x-ray photoelectron spectroscopy, we are justified in treating it as a constant. The shape of the edge will therefore correspond to the density of empty states near the Fermi energy broadened by the lifetime of the core hole. That is, to a first approximation it should look like an x-ray absorption edge. The transition density is thus given by the convolution of the conduction band density of states $N_c(E)$ and the distribution function of the initial core level $N_i(E)$, i.e.,

$$T(E) = \int_0^E N_c(E') N_i(E - eV + E\varphi_{\text{cath}} - E') \, dE'. \tag{5}$$

We have neglected the possible effects of many-electron dynamic screening of the core hole (Chang and Langreth, 1972).

Unfortunately the strong electron–electron interaction, necessary in order for the technique to be surface sensitive, has a less desirable effect. The sky above the sample is thick with secondary electrons, very few of which are ejected from core levels. To enhance the weak but relatively sharp features associated with the ionization edges, we can selectively amplify the higher frequency spectral components of the secondary electron spectrum by measuring its energy derivative. This is equivalent to weighting each Fourier component of the spectrum by its frequency.

Of the several possible means of measuring energy derivatives of the secondary electron spectrum, the most flexible and widely used is the potential modulation technique (Leder and Simpson, 1958). In this technique a small oscillation is superimposed on the pass band of the electron spectrometer. A phase-sensitive amplifier selects that portion of the current through the analyzer that varies at the modulation frequency or one of its higher harmonics. The nth harmonic amplitude is proportional to the nth derivative with respect to energy, broadened by a response function that depends on the oscillation amplitude (Houston and Park, 1972a).

In an early application of potential modulation to enhance Auger emission features in the secondary electron spectrum of beryllium, Harris (1968) recognized the beryllium K-shell ionization edge. Core level ionization edges were subsequently reported in the energy derivative of the secondary electron energy distribution of other materials (Ellis and Campbell, 1968; Bishop and Riviere, 1970). Unlike photoemission spectra, however, the ionization edges are generally much weaker than the far more abundant Auger peaks, since the ejected electrons are distributed over all energies lower than the edge.

To make much use of these features, they must be distinguished from the Auger spectrum in which they are immersed. This distinction depends on the fact that the ionization edges are fixed with respect to the incident electron energy. The Auger features, on the other hand, are fixed relative to the Fermi energy of the sample and are thus independent of the energy of the incident electrons. Two methods of exploiting this distinction have been employed to separate the ionization edges from the total energy distribution.

Using a modification of a scheme employed by Houston and Park (1970) to remove characteristic loss features from the Auger spectrum, Gerlach *et al.* (1970) oscillated the incident electron energy a few electron volts about its average value. Only those features in the secondary electron energy distribution that varied synchronously with the primary electron energy were detected. This amounts to subtracting secondary electron distributions taken at two slightly different incident electron energies. The result is the approximate derivative of the characteristic loss spectrum.

A second method of separating the characteristic loss spectrum has been described by Gerlach and Tipping (1971). The technique is to fix the pass band of the electron spectrometer at an arbitrary energy and sweep the energy of the primary electrons. The characteristic loss features, including of course the ionization edges, are swept across the pass band, whereas features that are fixed relative to the Fermi energy, such as Auger peaks, are not. To reduce background variations it is generally necessary to measure the second derivative of the spectrum. A characteristic ionization loss spectrum of $GdCo_5$ obtained by J. E. Houston of Sandia Laboratories using this technique is shown in Fig. 4.

The surface was cleaned by argon ion sputtering and electron bombardment annealing before the spectrum was taken. All of the gadolinium and cobalt levels within the energy range of the spectrum are easily identified with the exception of the cobalt L_1. In addition, oxygen and sodium contaminants are present despite the cleaning procedure. Samples cut from the same $GdCo_5$ ingot were examined by several other techniques, and as we shall see, they all indicate that oxygen is distributed throughout the sample.

Thus we see that it is indeed possible to obtain an inner-shell excitation spectrum of surface atoms with incident monoenergetic electrons rather than x-ray photons. The view of the surface is, however, quite different. Associated with each core level we obtain an edge that reflects the density of conduction band states near the Fermi energy. This information is complementary to the valence band spectrum obtained by XPS at small binding energies.

There is, however, a very important distinction. In principle the only limitation on the detail that can be detected in the density of states by a direct excitation technique such as XPS is the resolution of the instrument. This is not the case with indirect (core level) excitation techniques. The localized core electron overlaps only part of the conduction states. By using the core level as a window, we view a very local density of states. The

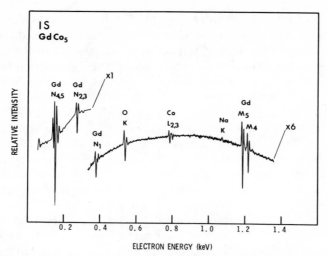

Fig. 4. Characteristic electron energy loss spectrum of a GdCo$_5$ sample obtained by J. E. Houston, Sandia Laboratories, using the ionization spectroscopy technique. A two-stage retarding-cylindrical mirror analyzer was set to pass electrons of fixed energy (100 eV) and the primary electron energy varied to move the characteristic loss features across the analyzer window. The second derivative of the analyzer current with respect to incident electron energy is plotted as a function of incident electron energy less the pass energy of the analyzer. The shape of the edges reflects the density of conduction band states broadened by the Auger lifetime of the core hole and instrumental factors. The surface was cleaned by sputtering with argon ions, but a pronounced oxygen edge is still evident. The Na K edge at 1072 eV apparently results from system contamination during earlier studies of Na adsorption. The spectrum was obtained in 7 min with a primary current of 150 μA. The resultion is $\sim$10 eV.

electronic structure of the local atomic environment (Haydock *et al.*, 1972) is at the root of understanding chemisorption that lies somewhere between the conventional band structure models of solid state physics and the molecular orbital view of the chemist. It is an inevitable consequence of the uncertainty principle that a localized view will be fuzzy. This fuzziness takes the form of broadening introduced by the finite lifetime of the core hole.

3. Soft X-Ray Appearance Potential Spectroscopy (SXAPS)

At about the time, a half century ago, that Robinson and his colleagues were measuring core electron binding energies by XPS, Richardson and Bazzoni (1921) were extending the Franck–Hertz concept to the measurement of "critical potentials" for the electron impact excitation of inner-shell electrons. The critical or "appearance" potential is the minimum potential that must be applied between an electron emitter and a sample to create a given core vacancy. The relation between the appearance potential and the

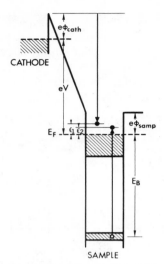

Fig. 5. Energy level diagram of the appearance potential experiment. An incident electron of energy $eV + e\varphi_{\text{cath}}$ may be captured in a state ε_1 above the Fermi energy exciting a core electron into a state $\varepsilon_2 = eV + e\varphi_{\text{cath}} - \varepsilon_1 - E_B$ above the Fermi energy. The threshold for excitation of the core electron is at $eV + e\varphi_{\text{cath}} = E_B$. The excitation probability above the threshold should vary as the self-convolution of the unfilled (conduction band) density of states broadened by the lifetime of the core hole. Core electron excitations are signaled in soft x-ray appearance potential spectroscopy by an increased emission of x-rays, in Auger electron appearance potential spectroscopy by an increase in secondary emission, and in disappearance potential spectroscopy by a decrease in the elastic scattering coefficient.

binding energy is illustrated by the energy level diagram in Fig. 5. The incident electron, which has an energy $eV + e\varphi_{\text{cath}}$ relative to the Fermi level of the sample, may be captured in a state ε_1 above the Fermi energy. If the energy it gives up is absorbed by a core electron, the core electron will be excited to a level ε_2 above the Fermi energy given by

$$eV + e\varphi_{\text{cath}} = E_B + \varepsilon_1 + \varepsilon_2. \tag{6}$$

There is a minimum potential, eV_{crit}, below which the incident electrons cannot inelastically scatter from that particular electron shell, given by

$$eV_{\text{crit}} = E_B - e\varphi_{\text{cath}}. \tag{7}$$

The excitation of an atomic core level is signaled in several ways. The total x-ray emission increases above the threshold since a new channel for producing characteristic x-rays is opened; the total secondary emission of electrons also increases above the threshold as a result of Auger recombination of the core hole; and, in direct analogy to the Franck–Hertz experiment, the elastic scattering coefficient decreases because a new mode of inelastic scattering is available. As we shall see, the appearance potentials can be determined from each of these signals.

The approach taken by Richardson and Bazzoni (1921) was to measure the current from a metal photocathode exposed to x-rays from an electron-bombarded sample. A plot of the photocurrent as a function of the potential applied between a graphite sample and tungsten filament exhibited a distinct break in slope at the threshold potential for excitation of the carbon K shell. They determined the K-shell binding to be 286 eV compared to the presently accepted value of 284 eV.

As in the case of ionization spectroscopy, the shape of an excitation edge is related to the density of conduction band states, but in a less obvious way. Assuming constant oscillator strengths, the excitation probability is proportional to the integral product of the initial- and final-state distributions, where the initial states are filled core levels. The final states must take into account all possible combinations of energies ε_1 and ε_2 for the two electrons (incident and excited core electron) allowed by the conservation of energy, as expressed by Eq. (6). The density of final states for the two electrons is thus given by the self-convolution of the density of the conduction band states for one electron $N_c(E)$, i.e.,

$$N_{2c}(E) = \int_0^E N_c(E') N_c(E - E') dE'. \tag{8}$$

The transition probability then is the integral product of $N_{2c}(E)$ with the distribution function describing the initial core level. As we have pointed out previously, a core hole has a finite lifetime τ, and the core level must therefore be described by a distribution function $N_i(E)$ having a width $\hbar/\tau$. The transition density function (Dev and Brinkman, 1970) is given by

$$T(E) = \int_0^E N_{2c}(E') N_i(E + E_B - E') dE'. \tag{9}$$

According to this simple picture, the edge should have the shape of the self-convolution of the density of conduction band states, broadened by the core level window. As in the case of ionization spectroscopy, this broadening is a necessary consequence of the local view of the density of states.

It is not necessary, of course, that the incident electron be scattered from a core electron for an x-ray photon to be produced. Radiative capture of the incident electrons in states above the Fermi energy creates a large bremsstrahlung background on which the core level excitation edges can only infrequently be observed. The advantages of potential modulation differentiation in the suppression of slowly varying backgrounds were apparent with respect to ionization spectroscopy in the last section and will be again when we discuss electron-excited Auger spectroscopy in the next section. The bremsstrahlung background, on which the excitation probabilities of the core levels are superimposed, should be particularly easy to suppress because of its nearly linear dependence on energy.

This fairly obvious fact led Park *et al.* (1970) to construct a soft x-ray appearance potential spectrometer that used an electron multiplier to detect the total x-ray emission of an electron-bombarded surface. The derivative of the x-ray emission as a function of sample potential is obtained by super-imposing a small oscillation on the potential and detecting that portion of the detector current that varies at the modulation frequency. The multiplier was subsequently replaced by a simple photocathode by Park and Houston (1971). This represents the simplest existing technique for surface chemical analysis, being no more complicated than an ion gauge. Moreover, it has the highest resolution of any core level electron spectrometer, since the resolution is limited only by the energy spread of the incident electrons, and this can be kept below 0.25 eV.

The high resolution attained in SXAPS makes it an attractive technique for the study of the local density of states. Nilsson and Kanski (1973) made a detailed comparison of unfolded appearance potential spectra of simple metals with theoretical one-electron densities of states and found excellent agreement for Al and Be. They had earlier demonstrated good correlation between appearance potential spectra of the Al $L_{2,3}$ edge and x-ray adsorption measurements (Nilsson and Kanski, 1972). Park and Houston (1972) demonstrated that the L_3, L_2 peak widths of the 3d transition metals show a systematic decrease with Z corresponding to the filling of the 3d band.

In some materials, however, notably boron and graphite, the simple one-electron excitation model described by Eq. (9) is clearly inadequate. In these materials, dynamic screening of the suddenly created 1s core hole results in a resonant coupling to plasmons (Langreth, 1971; Laramore, 1972). The effect is to produce a series of jumps in the excitation probability, separated by the plasmon energy (Houston and Park, 1972b). Many-electron effects of this sort are beyond the scope of this chapter, but they constitute a complication in all core level spectroscopies.

For elemental analysis the kind of resolution required for band structure studies is unnecessary, and it is convenient to increase the modulation and measure the second derivative of the x-ray emission. The second-derivative spectrum of a $GdCo_5$ alloy surface, obtained by D. G. Schreiner, Sandia Laboratories, is shown in Fig. 6. It is possible to identify all of the gadolinium and cobalt levels except those at very low energies. Traces of common surface impurities such as oxygen and carbon are detectable, although the surface had been cleaned by sputtering with 1 keV neon ions. The Mo $M_{4,5}$ spectrum results from electrons striking the Mo sample holder.

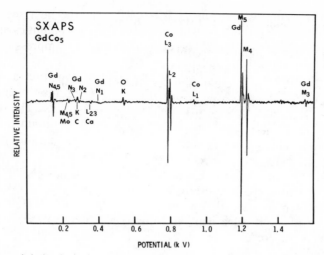

Fig. 6. Second-derivative soft x-ray appearance potential spectrum of $GdCo_5$ obtained by D. G. Schreiner, Sandia Laboratories. The second derivative of the current from a gold photo-cathode exposed to the sample surface during electron bombardment is plotted as a function of the bombarding potential. The surface was cleaned by sputtering with neon ions, but the oxygen K edge is still evident. Smaller amounts of Ca and carbon are also detectable. The Mo $M_{4,5}$ edge results from electrons striking the molybdenum sample holder. The shape of the excitation edges is related to the self-convolution of the conduction band density of states. The spectrum was obtained in 12 min with a primary current of 4 mA. The resolution is ~ 6 eV.

It appears, however, that there are interesting limitations to the use of SXAPS for elemental identification, although reports of insensitivity to specific elements such as copper (Tracy, 1971) and silicon (Tracy, 1972) have more frequently than not proven to be premature (Park and Houston, 1973b). Attempts to generalize this supposed insensitivity to classes of materials such as low Z elements (Tracy, 1971), semiconductors, insulators, 4d transition metals, and electronegative impurities (Tracy, 1972) have been even less successful. In fact, beryllium at $Z = 4$ (Nilsson and Kanski, 1973) is observed with about the same sensitivity as uranium at $Z = 92$ (Park and Houston, 1973c); silicon (Park and Houston, 1973b) and boron (Houston and Park, 1971) among the elemental semiconductors give good K spectra; $Sc_2 O_3$ (Park and Houston, 1973a) is an example of an insulator that produces a strong spectrum, as do electronegative impurities such as sulfur and oxygen (Park and Houston 1973b).

Gold and palladium, on the other hand, are examples of elements that do not give detectable spectra by existing appearance potential techniques. It appears that in the energy range of interest for surface studies (< 2000 eV) the matrix elements connecting the core levels of these elements to unoccupied states vanish for states near the Fermi energy. The reasons for this behavior should provide an interesting subject for scientific inquiry.

Apearance potential spectroscopy, of course, has roughly the same depth sensitivity as XPS. In the appearance potential case the sampling depth is determined by the inelastic damping length of the incident electrons rather than the ejected core electrons as in XPS. Thus the sampling depth of the L_3 level of iron (706 eV) should be almost identical for SXAPS and XPS using the Al Kα line (1.49 keV). However, in those cases where a careful determination of binding energies has been made by the appearance potential technique—namely, the L levels of the 3d transition metals (Park and Houston, 1972), the N and O levels of thorium (Redhead and Richardson, 1972), and uranium (Park and Houston, 1973c)—the appearance potential values have been consistently lower than tabulated binding energies based on XPS measurements (Bearden and Burr, 1967). It seems likely that much of this discrepancy results from differences in the state of cleanliness of the surfaces studied.

Quite apart from the question of cleanliness, however, atoms in the outer-most layer of a solid reside in a different chemical environment from their bulk counterparts simply by virtue of their reduced coordination. The resulting modification of atomic potentials should produce a "surface chemical shift" in the core levels. Direct experimental evidence of this shift has been obtained for the $2p_{3/2}$ levels of Ti, Cr, and Ni (Houston et al., 1973). A direct comparison was made of the electron-excited soft x-ray appearance potentials, which are surface sensitive, and photon-excited Auger electron appearance potentials, which are more characteristic of bulk atoms. A single tungsten emitter served as a reference for both measurements. The "surface chemical shifts" thus obtained were in reasonable agreement with predictions based on a muffin-tin approximation of over-lapping one-electron potentials.

4. APPEARANCES AND DISAPPEARANCES

We should not leave the subject of appearance potentials without at least a brief comment on the use of signals other than soft x-rays to mark the excitation thresholds. Indeed, in the energy range of interest for surface studies, radiative recombination of a core hole is a relatively rare event compared to Auger processes. It is not surprising therefore that Richardson (1928) observed that at every potential where there is an increased emission of soft x-rays, there is a corresponding increase in the emission of secondary electrons. These Auger electron appearance potentials were used by Houston and Park (1970) in a cross-correlation experiment to retrieve the Auger spectrum resulting from the excitation of a specific core level. Gerlach (1971) demonstrated that the appearance potentials of disordered materials can be obtained by simply plotting the derivative of the total secondary emission of an electron-bombarded surface in a manner analogous to SXAPS. For ordered crystalline solids, however, the secondary emission background exhibits a great deal of structure that is unrelated to the excitation of core holes. This structure, which is due to electron diffraction, damps out at higher energies, however, and the 2p spectrum of Cr was obtained by this method (Houston and Park, 1972c).

It is interesting to note that SXAPS is limited at higher energies by a monotonically increasing shot noise contribution from the bremsstahlung background. The secondary emission background does not continue to increase with energy, however, suggesting that Auger electron appearance potential spectroscopy (AEAPS) may prove to be of great value at higher energies.

Although the total secondary emission increases at an excitation threshold, the elastic scattering coefficient decreases because of the availability of a new channel for inelastic scattering. This has been used by Kirschner and Staib (1973) in a direct extension of the Franck–Hertz experiment which they antithetically refer to as "disappearance potential spectroscopy." The diffraction problems that plague AEAPS, however, are even worse if one confines himself to elastically scattered electrons. The diffraction variations in the total secondary emission are, in fact, almost solely the result of the elastic component. There is, of course, information on the geometric arrangement of atoms in these variations that raises the intriguing possibility of simultaneous determinations of crystallography and composition by separately recording the elastic and inelastic components. There are, in any case, many new avenues yet to be explored.

B. Deexcitation Techniques

The appearance potential techniques are indifferent to the energies of the characteristic x-rays or Auger electrons emitted during the reorganization of an excited atom. Their existence is inferred solely from their contribution to the total number of emitted photons or electrons. Only the energies of the incident electrons are analyzed. The game can also be played the other way around, using any convenient method of creating the initial vacancy and analyzing the energies of the photons or electrons emitted by the atom as it convulses its way back to the ground state.

The most obvious distinction between the excitation and deexcitation spectra is the relative complexity of the latter. We are no longer dealing just with the atomic levels but with their term differences.

1. ELECTRON-EXCITED AUGER ELECTRON SPECTROSCOPY (EEAES)

The first practical Auger spectrometer seems to have been constructed by Steinhardt and Serfass (1951) using x-ray excitation. This early venture into what is now generally called ESCA (electron spectroscopy for chemical analysis) did not receive the attention it deserved. The electron background above which the Auger peaks must be detected in the x-ray-excited spectrum is quite manageable, but for electron bombardment excitation it is another matter. Auger peaks were first identified in secondary electron energy distributions by Lander (1953), despite a voluminous literature on secondary electron distributions (McKay, 1948) prior to that time. Only the most intense Auger transitions could be detected amongst the hail of secondary electrons that result from the strong interaction of the incident electron beam with the valence electron fluid of the solid. In the first of many applications of potential modulation to surface analysis, Harris (1968) demonstrated the remarkable enhancement of the relatively sharp Auger features that can be achieved by differentiation.

The acceptance of Auger electron spectroscopy was facilitated by the fact that many surface physicists already had spherical grid low-energy electron diffraction systems that could be converted to retarding potential energy analyzers by the addition of external electronics (Weber and Peria, 1967). Because of the obvious advantages of coupling low-energy electron diffraction (LEED) with Auger electron spectroscopy, the retarding analyzer remains an important instrument. For applications where LEED would be irrelevant, however, the cylindrical mirror analyzer (Sar-el, 1967), with its relatively high efficiency, is now widely used. Spectra can be obtained in a much shorter time than is possible with retarding analyzers, but claims of increased sensitivity are difficult to verify. As with all deflection analyzers, the resolution of the instrument varies with the energy. This distorts the spectrum, but at low energies it approximately compensates for the rapidly varying background produced by the true inelastic peak and gives a more aesthetic spectrum.

The acquisition of the spectrum may, however, be far simpler than its interpretation. The L Auger spectrum alone consists of hundreds of lines for a heavy element. These lines can be thought of in terms of a transition of holes. Thus, for example, the designation $K L_1 L_2$ refers to an initial vacancy in the K shell that undergoes a transition to holes in the L_1 and L_2 shells plus an Auger electron.

The theory of the KLL transitions has been treated more extensively than that of any other series. In electron-excited Auger spectroscopy of surfaces, however, the incident electrons usually have an energy of 3 keV or less. This is less than the K-shell binding energy of elements above $Z = 17$. For elements with $Z < 18$ the coupling can be taken as pure LS (Burhop and Asaad, 1972), which results in five lines as compared to the nine lines predicted by intermediate coupling. This has been confirmed experimentally by Korber and Mehlhorn (1966) for $Ne(Z = 10)$ and by Fahlman *et al.* (1966) for $Mg(Z = 12)$. The energies of the lines are in excellent agreement with calculations of Asaad (1970).

The L Auger spectrum is far more complex than the K spectrum, and theoretical calculations taking into account relativistic and intermediate coupling effects have not been carried out. The result is that serious ambiguities arise in the identification of L Auger peaks. Bergström and Hill (1954) proposed that the energy of a transition abc be calculated by the formula

$$E_{abc}(Z) = E_a(Z) - E_b(Z) - E_c(Z) - \Delta Z \{E_c(Z+1) - E_c(Z)\}. \quad (10)$$

The rationalization for this equation is that the Auger electron is ejected from the c state by external conversion of a photon produced by the transition of the initial a vacancy to b. Because of the b vacancy, the binding energy of the c electron is not that of a neutral atom, but of an atom with an apparent nuclear charge of $Z + \Delta Z$. ΔZ is found empirically to be between 0.5 and 1. Chung and Jenkins (1970) have proposed a modification of Eq. (10) that seems to require a smaller variation of ΔZ.

Except for light elements, however, elemental identification is generally based on matching spectra against "standard" plots taken from samples of known composition. For example, in the spectrum of $GdCo_5$ shown in Fig. 7, obtained by R. G. Musket, Sandia Laboratories, it is generally possible to identify spectral groups but not individual features within those groups. At low energies, however, not even spectral groups can be specified, and identification of features as belonging to Gd or Co depends on comparison with spectra of the pure metals. The surface was cleaned by argon ion sputtering before taking the spectrum, and the KLL spectrum of argon imbedded in the lattice is clearly detectable. Oxygen is once again present in spite of the sputtering.

The relationship between the line shapes and the density of valence band states is shown schematically in Fig. 8 for a core level valence–valence transition. The kinetic energy of the ejected electron will depend on the positions of the two valence band holes. If we assume the core hole recombination is with an electron in a state ε_1 in the valence band and the Auger electron is ejected from a state ε_2, also in the valence band, then the kinetic energy is given by

$$E_K + e\varphi_{spec} = E_B - \varepsilon_1 - \varepsilon_2. \quad (11)$$

Thus the energy distribution of Auger electrons must take into account all

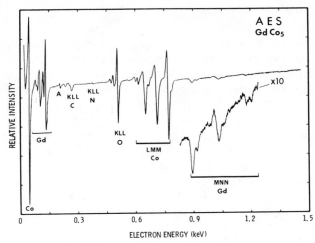

Fig. 7. Electron-excited Auger electron spectrum of GdCo$_5$ obtained by R. G. Musket, Sandia Laboratories. The surface was cleaned by argon ion sputtering prior to taking the spectrum. The derivative of the secondary electron energy distribution was obtained by sweeping the pass energy of a cylindrical mirror analyzer as the surface was bombarded with 3 keV electrons at $\sim 100\,\mu A$. The KLL Auger transition is due to argon trapped in the lattice during the sputter cleaning. The oxygen spectrum is relatively strong compared to most of the other techniques, and the gadolinium spectrum is relatively weak. Nitrogen and carbon are also detectable. The $\times 1$ spectrum was taken in 25 min with a modulation of 2.5 eV p-p. The resolution decreases with energy.

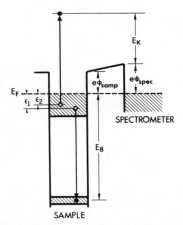

Fig. 8. Energy level diagram of the Auger electron emission experiment. The initial state is taken as including a vacancy in a core hole. The energy liberated when the core hole recombines with a valence electron is transferred to another valence electron that may be ejected into the vacuum with an energy $E_B - \varepsilon_1 - \varepsilon_2$. The measured kinetic energy must be corrected for the work function of the spectrometer, $e\varphi_{\mathrm{spec}}$. The shape of the core level–valence–valence Auger line reflects the self-convolution of the valence band density of states broadened by the lifetime of the core hole. The maximum kinetic energy of an Auger electron corresponds to the case where both final state holes are at the top of the valence band ($\varepsilon_1, \varepsilon_2 = 0$).

possible combinations of energies ε_1 and ε_2 allowed by the conservation of energy, as expressed in Eq. (11). The transition density is therefore given, in this simple picture, by the self-convolution of the valence band density of states (Lander, 1953; Amelio and Scheibner, 1968) broadened by the width of the core level, i.e.,

$$T(E) = \int_0^E N_{2v}(E') N_i(E + E_B - E') dE', \qquad (12)$$

where

$$N_{2v}(E) = \int_0^E N_v(E') N_v(E - E') dE'. \qquad (13)$$

Comparing Eqs. (12) and (9) we see that AES provides information about the valence band complementary to that provided by APS for the conduction band. A systematic comparison of line shapes obtained by both techniques using the same core level windows would seem to be a rewarding project. Until recently, however, the emphasis in EEAES has been on contamination monitoring (Taylor, 1969), and few high resolution studies of line shapes have been attempted. In AES, as in all core level spectroscopies, many-electron effects can be expected to complicate the interpretation of line shapes (Natta and Joyes, 1970).

Amelio (1970) did report reasonable agreement between the deconvoluted silicon $L_{2,3}$ VV transition and calculated densities of states for silicon. He did not, however, account for instrumental broadening or characteristic loss effects. Mularie and Peria (1971) showed that the elastic scattering peak and its associated characteristic loss features provide a satisfactory instrument response function to correct for both the instrument and inelastic interactions.

Sickafus (1973) correlated EEAES measurements of the sulfur $L_{2,3}$ VV transition for a (110)Ni–$c(2 \times 2)$S surface with measurements by Hagstrum and Becker (1971) using ion neutralization Auger spectroscopy (INS). There is, in fact, no a priori reason to expect agreement. INS is a direct excitation technique and sees a valence band structure averaged over all sites, both Ni and S. EEAES, on the other hand, takes a very local view of the valence band, as we have discussed elsewhere.

Perhaps the highest resolution EEAES spectra (0.2 eV) are the $L_{2,3}$ VV Auger spectrum of aluminium and the $M_{4,5}$ VV Auger spectrum of silver obtained by Powell (1973). He concludes that there is no simple relationship between the silver spectrum and the valence-band density of states. He attributes this behavior to final-states interactions.

There have been numerous reports of chemical shifts in the positions of Auger peaks (Fahlman et al., 1966; Haas and Grant, 1970; Coad and

Riviéré, 1971; Szalkowski and Somorjai, 1972). While it is clear that something in the spectrum has shifted, it should not be supposed that the binding energy of the core electron has necessarily changed. Changes in the shape of the valence-band density of states will also produce an apparent shift. They are, however, chemical changes and as such may be useful in analysis.

Some progress has been made in efforts to make Auger spectroscopy more quantitative (Meyer and Vrakking, 1972; Houston, 1973), but, as with all surface analytical techniques, the situation is far from satisfactory in spite of optimistic pronouncements (Palmberg, 1973). Gallon (1972), for example, has considered the effect of ionization by backscattered secondary electrons. He concludes that for the $L_{2,3}$ VV transition in pure silicon, about 20% of the ionization is by backscattered electrons. Gerlach and DuCharme (1972) computed similar corrections in measurements of ionization cross sections of C on W and conclude that as much as 25% of the K-shell excitation is by backscattered secondaries. In a more direct determination, Tarng and Wehner (1973) find that about 40% of the Mo Auger signal from thin layers on a W substrate is generated by backscattering.

This problem is much less severe in APS and IS. Only elastically scattered secondaries can contribute any enhancement to the excitation edges, and these are too insignificant to cause much concern. In the Auger case, however, inelastic secondaries may well be more efficient than primary electrons for the excitation of lower energy shells (Houston and Park, 1969). The importance of such problems as surface segregation in alloys, however, should provide ample motivation for the solution of these problems.

2. Soft X-Ray Emission

The classic method of soft x-ray band spectra, usually regarded as a bulk analytical technique, becomes surface sensitive as the incident electron energy is reduced to near threshold values. Such studies were originally undertaken to avoid the distortion introduced by self-absorption (Liefeld, 1968), but it is clear that elimination of self-absorption is equivalent to probing only the outermost layer of a solid. Unfortunately, the excitation probability also approaches zero near the threshold and soft x-ray analysis, difficult under normal excitation conditions, becomes even more exacting.

A great deal could be learned by contrasting soft x-ray band spectra with Auger electron spectra taken under the same conditions. Radiative recombination of the initial core vacancy with a valence electron involves only a single hole in the valence band and, to a first approximation, the transition density goes as

$$T(E) = \int_0^E N_v(E') \, N_i(E + E_B - E') \, dE'. \tag{14}$$

Thus the local one-electron density of valence band states is obtained without unfolding. The spectrum will, of course, still be complicated by dynamic screening effects. The transition probabilities are further complicated by selection rules governing the initial and final states, since even in the valence and conduction bands the admixture of the various wave function symmetries is not complete. It is therefore necessary to combine the spectra of two or more initial core levels of different symmetry to obtain the occupancy of the valence band.

There is an alternative to the near-threshold excitation as a means of making electron-excited soft x-ray band spectra sensitive to the surface region. The penetration depth of high energy electrons can be reduced by using grazing incidence (Cuthill *et al.*, 1970). A glancing incidence reflection high energy electron diffraction system (RHEED) was, in fact, combined with a soft x-ray spectrometer by Sewell and Cohen (1967) at about the same time that Harris published his initial description of an electron-excited Auger spectrometer (Harris, 1968). This combined RHEED and x-ray analysis system permits simultaneous determination of structure and composition. The resolution of the instrument does not allow x-ray line shape analysis, however, and the sensitivity is perhaps an order of magnitude less than AES for light elements.

The field of surface soft x-ray band spectra remains largely unexplored. Its principal practitioners are not oriented toward surface studies, and surface scientists have shown little inclination to undertake such demanding experiments.

3. ION-INDUCED X-RAYS

In studying the emission of characteristic soft x-rays, it is necessary to contend with a background of bremsstrahlung photons produced when incident electrons are radiatively captured in unoccupied states above the Fermi level. It is therefore attractive to consider other ways of creating the initial core vacancies, since in principle they can be created by any projectile of energy greater than the binding energies of the core electrons.

Protons and alpha particles are fairly obvious candidates for the probing particle. The production of bremsstrahlung by the collision of these particles with solids is generally negligible (Merzbacher and Lewis, 1958), but unfortunately the x-ray yield is also quite small as compared to electrons. It is thus no longer practical to use dispersive analyzers. Measurements are made instead with solid state detectors.

There have been great strides made in recent years in solid state energy analysis. Unlike dispersive analyzers, which actually discriminate on the basis of momentum rather than energy, solid state energy analyzers are not confined to measurements from point sources. However, a state-of-the-art resolution of 145 eV is not likely to impress soft x-ray and electron spectroscopists accustomed to thinking in terms of tenths of electron volts. What is sacrificed of course is any sensitivity to chemical environment.

For elemental analysis the simplicity of the x-ray emission spectrum as compared to the Auger electron spectrum compensates to a large extent for the lower resolution. Thus Cr and Fe are resolved about as easily in the proton-induced x-ray (PIX) spectrum of stainless steel (Musket and Bauer, 1972) as in the Auger electron spectrum (Tracy and Palmberg, 1969), and the monolayer sensitivity is at least comparable (Van der Weg *et al.*, 1973). A PIX spectrum, taken by R. G. Musket and W. Bauer, Sandia Laboratories, with 350 keV protons is shown in Fig 9 for comparison with spectra taken by other probes. The K_α line of the omnipresent oxygen contaminant is clear. The carbon K_α line represents the practical low-energy limit with existing technology.

The principal advantage of PIX may well be that it can be combined with Rutherford backscattering for the study of surface films (Bauer and Musket, 1973). We will encounter the latter technique again when we discuss techniques based on nuclear mass (Section III.A.2).

Smaller sampling depths, and in some cases much higher x-ray yields, can be achieved with heavy ions (Saris, 1972). The results, however, can be quite

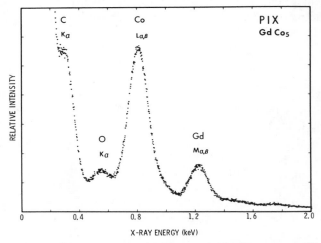

Fig. 9. Proton-induced x-ray spectrum of GdCo$_5$ obtained by W. Bauer and R. G. Musket, Sandia Laboratories. The number of x-ray photons per unit energy resulting from bombardment of the surface with 350 keV protons was measured with a Si(Li) detector with ~ 145 eV resolution. The carbon Kα line represents the practical low-energy limit with existing techniques. The rise in background at low energies is a noise effect. The incident beam current was 150 nA. The average sampling depth is somewhat greater than for the electron spectroscopies.

complex. If there is interpenetration of the electronic shells of the incident ions with atoms in the target, short-lived quasimolecules are formed that can greatly alter the excitation of core levels. The spectrum, of course, includes x-rays from the probing ion as well as from the sample. By proper choice of the atomic number and energy of the incident ion, remarkable sensitivities can be attained for specific elements. This is a new and still poorly understood area, but it is a potentially powerful tool for probing the chemistry of surfaces.

III. NUCLEAR MASS

The mass of a surface atom is determined either from the recoil momentum given an incident ion or by bodily removing surface atoms for analysis by a mass spectrometer. The latter approach is clearly destructive, and so indeed is the recoil method, but the destruction can generally be kept within very tolerable limits by either approach.

Much of the discussion of core level spectroscopies dealt not with simple elemental identification but with information on the chemical environment that can be extracted from secondary features of the spectra, such as chemical shifts in the positions of core levels. It is clear that things are going to be very different when the analysis is based on the masses of surface atoms. The mass defect of a chemical bond is, after all, generally beyond our power to resolve. There are, however, other chemical effects that manifest themselves in the mass analysis techniques.

A. Ion Backscattering

The collision of energetic ions with a surface has several consequences. In addition to the production of x-rays characteristic of both the target and projectile, the incident ions may impart sufficient momentum to surface atoms to cause sputtering. Some of the incident ions will also recoil from the surface either as ions or neutrals. The energies of these backscattered particles can generally be treated within the framework of classical mechanics —changes in their internal energy, for the most part, being a small perturbation.

1. BINARY SCATTERING OF NOBLE GAS IONS

When an ion approaches a metal surface, the most likely event is that it will be neutralized by an electron tunneling from the solid (Hagstrum, 1954). Some ions, however, survive the encounter and are backscattered as ions.

Smith (1967) demonstrated experimentally that for noble gas ions in the energy range of 0.5 to 3 keV, this backscattering is satisfactorily described as a simple elastic binary collision with a surface atom. Thus the masses of the atoms on a surface can be determined from the recoil momentum of the scattered ions, as shown schematically in Fig. 10.

From the classical conservation laws we can write

$$\frac{P_0^2}{2M_0} = \frac{P_S^2}{2M_0} + \frac{P_T^2}{2M_T}, \tag{15}$$

and

$$\mathbf{P}_0 = \mathbf{P}_S + \mathbf{P}_T, \tag{16}$$

where $\mathbf{P}_0$ is the momentum of the incident ion of mass M_0, $\mathbf{P}_S$ is the recoil momentum of the scattered ion, and $\mathbf{P}_T$ is the momentum imparted to a target atom of mass M_T. Eliminating the momentum of the target atom from Eqs. (15) and (16), we obtain a quadratic relationship between the momentum of the incident and scattered ion given by

$$P_0^2(M_T - M_0) + 2M_0 P_0 P_S \cos\theta - P_S^2(M_T + M_0) = 0, \tag{17}$$

where θ is the scattering angle. For $\theta = 90°$ the cross term drops out, and Eq. (17) assumes a particularly simple form:

$$\frac{P_S^2}{P_0^2} = \frac{E_S}{E_0} = \frac{M_T - M_0}{M_T + M_0}. \tag{18}$$

For this reason, and apparently no other, spectrometers to study binary scattering seem invariably to be constructed with a fixed 90° scattering angle. If $M_0 > M_T$, the scattering angle will always be greater than 90°.

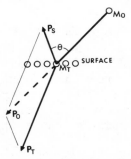

Fig. 10. Diagram of the binary scattering experiment. The recoil momentum P_S of an ion of mass M and initial momentum P_0 incident on a surface is found experimentally to be satisfactorily described by a simple elastic binary collision with a single surface atom of mass M_T. Measurements are usually made at a fixed scattering angle of 90°, in which case the ratio of the scattered ion energy to the incident energy is $E_S/E_0 = (M_T - M_0/M_T + M_0)$. If $M_0 > M_T$, the scattering angle θ will be greater than 90°.

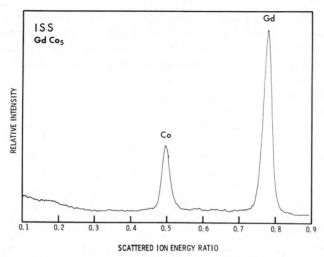

Fig. 11. Binary scattering spectrum of a GdCo$_5$ surface obtained by G. C. Nelson, Sandia Laboratories, with the ion scattering spectrometer. The spectrum was obtained with Ne$^+$ using a 90° scattering angle. The peak due to scattering from ^{59}Co is at the expected ratio E_S/E_0 = 0.495. Gadolinium has isotopes of nearly equal abundance at masses 155, 156, 157, 158, and 160 that are not resolved, but the maximum at E_S/E_0 = 0.78 is in good agreement with the average atomic weight of about 157. Atoms lighter than the ^{20}Ne$^+$ probe ions, such as ^{12}C and ^{16}O, cannot be detected, but presumably could be by using ^{4}He$^+$ as the probe. The spectrum was taken in 10 min at an incident ion current of 150 nA.

The energy spectrum of ^{20}Ne$^+$ scattered at 90° from a sputter-cleaned GdCo$_5$ surface, obtained by G. C. Nelson, Sandia Laboratories, is shown in Fig. 11. Atoms lighter than the ^{20}Ne$^+$ probe ions, such as carbon and oxygen, cannot be detected, but presumably could be by using ^{4}He$^+$ as the probe. The peak due to scattering from ^{59}Co occurs at E_S/E_0 = 0.495, as predicted by Eq. (17). Gadolinium has isotopes of roughly equal abundance at masses 155, 156, 157, 158, and 160 that are not resolved, but the maximum at E_S/E_0 = 0.78 is in good agreement with the expected position for the average atomic weight of about 157. Brongersma and Mul (1973) have discussed the factors that limit the achievable resolution, such as thermal vibrations of the surface atoms and changes in the internal energies of the colliding atoms.

We have identified the spectrum in Fig. 11 by the acronym ISS for ion scattering spectroscopy (Goff, 1972), which is fairly simple but not too informative. At the other extreme is the acronym NIRMS favored by Brongersma and Mul (1973), which stands for noble ion reflection mass spectroscopy.

Regardless of its name, the sampling depth of the binary scattering technique is confined almost entirely to the outermost layer of atoms (Brongersma and Mul, 1972). This permits some structural information to be deduced from shadowing effects. For example, in one of the first applications of the technique, Smith (1967) showed that He$^+$ scattering from carbon monoxide adsorbed on nickel gave little evidence of carbon, from which he concluded that the CO molecule is bound to the surface through the carbon atom with the oxygen exposed. The principal limitation to this kind of study is imposed by sputtering, which exposes the underlying atoms.

Sputtering can be an asset for some purposes, however, since by increasing the incident ion current, a profile of the composition into the bulk is obtained as the beam erodes the surface away. There is evidence in fact that in the absence of damage produced by the incident ions, the binary scattering relation does not hold (Heiland et al., 1973). There is, after all, no a priori reason to expect that the bonds of a surface atom to its neighbors can be ignored in ion scattering. Heiland et al. (1973) find that for a well-annealed single crystal nickel surface, the backscattering cannot, in general, be interpreted in terms of simple binary scattering. As the incident ions stir the surface, however, the binary peaks return.

The binary scattering technique has been covered in several review articles (see, for example, Smith, 1971; Armour and Carter, 1972; Carter, 1973). The amount of effort expended on binary scattering is still quite small, as compared to AES for example, and further improvements can be expected.

2. RUTHERFORD SCATTERING

At energies much greater than those employed in binary scattering studies, ions may penetrate many atomic layers into the solid before being backscattered. It is generally assumed that, before and after the backscattering event, the ion gives up energy uniformly to electronic excitation. The stopping power is therefore a characteristic of the material. This has recently been shown to be a powerful technique for the analysis of surface films (Meyer et al., 1970; Picraux and Vook, 1971) using MeV He ions as the incident particles and solid state detectors for the energy analysis. The resolution is about 100 Å, which is much greater than the sampling depths with which we are concerned in this paper.

By going to lower ion energies and using a dispersive analyzer, however, Powers and Whaling (1962) obtained much better depth resolution. This has been carried a step further by van Wijngaarden et al. (1971), who employed 50–100 keV protons and an electrostatic analyzer to detect oxide layers as thin as 6 Å on nickel surfaces. This brings Rutherford scattering into the same range as many of the other techniques for surface analysis. A Rutherford backscattering spectrum taken from the work of van Wijngaarden et al.

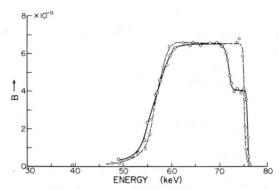

Fig. 12. Backscattered energy distribution obtained by van Wijngaarden *et al.* (1971) for 80 keV protons incident on a 350 Å Ni film. The circles (– – –) are for the clean nickel film. The squares (——) are for the same film after oxidation to a depth of ∼30 Å.

(1971) is shown in Fig. 12. It compares the spectrum of a nickel film before and after oxidation.

Combined with proton-induced x-ray analysis (Section II.B.3), Rutherford scattering provides a powerful method of studying very thin films.

B. Removal of Surface Atoms

All of the techniques that have been discussed unavoidably inflict some damage on the surface in the act of analyzing it. The composition can also be determined by deliberately destroying the surface and analyzing the debris. There are, of course, many ways of removing atoms or ions from a surface for analysis in a mass spectrometer. Some, such as thermal evaporation and electron-induced desorption, are too selective to provide complete information on surface composition, although they do provide a spectroscopy of binding states. Sputter removal and field evaporation, however, aided by very significant improvements in mass analysis techniques, have been used as the basis of surface analytical techniques with quite remarkable sensitivity characteristics.

1. SECONDARY ION MASS SPECTROSCOPY (SIMS)

The reorganization of a lattice following the impact of an energetic ion may result in an atom, or even a cluster of atoms, being ejected from the surface. Some of these ejected particles will be charged. The method of secondary ion mass spectroscopy (SIMS) consists of determining the mass-to-charge ratio of these sputtered ions by conventional mass analysis techniques—usually a quadrupole mass filter. Sputtered ions have been used to study the composition of bulk materials and thin films for many years (Honig, 1958). Until recently, however, it was not generally regarded as an appropriate method for the study of surface layers because of the rapid sputtering rates. Benninghoven (1970) has demonstrated, however, that by increasing the area covered by the ion beam, the primary ion current density can be reduced to a point where the monolayer removal time exceeds the monolayer contamination time in an ultrahigh vacuum system. Benninghoven terms this "static SIMS." It should be noted that for the study of chemisorbed molecules such as CO, which have large cross sections for electron-induced desorption, static SIMS is less destructive than electron bombardment techniques such as EEAES and SXAPS. The static method has been quite thoroughly reviewed by Benninghoven (1973).

The critical factor affecting the amount of damage that must be inflicted to achieve a given sensitivity is the yield of sputtered ions. The probability that a sputtered particle will be ionized exhibits an astonishing variation, depending on the chemical state of the surface. The yield of metal ions from an oxidized metal surface, for example, is typically 10^3 greater than for a clean surface. This is exciting information if one is seeking clues to the chemical state of surface atoms, but it poses grave problems for those attempting quantitative determinations of elemental concentrations.

Attempts to calculate ion yields have taken two antagonistic approaches. A model applicable to inert gas ion sputtering of metals (Schroeer et al., 1973) assumes that atoms are always sputtered as neutrals and are subsequently ionized by transitions of the sputtered atom's valence electrons to the top of the conduction band while the atom is still in the influence of the solid. The model therefore depends critically on the ionization potential of the atom. The ionization probability in this model goes as

$$P_1 \propto (v^n/\Delta E^{n+2}),\qquad(19)$$

where v is the velocity of the sputtered atom, ΔE is the ionization energy less the work function, and n is an adjustable parameter the authors find to be about 2.5.

The opposite point of view is that the atoms are sputtered as ions but may be neutralized by an electron tunneling from the solid (Anderson, 1970; Werner, 1969). This does not involve the ionization potential at all and predicts the neutralization probability will go as

$$P_N \propto \exp(-e\varphi/kT),\qquad(20)$$

where φ is the work function. It is a tribute to mathematical physics that qualitative agreement with experiment is reported for both models.

An interesting example of the effect of the matrix is that the ionic yield of metals present as dopants in semiconductors can be 10^2 times greater than for the pure metal (Werner and de Grefte, 1973). The yield of metal ions is a factor of 10^3 greater for oxides. Werner (1969) therefore proposed that O_2^+ be used as the bombarding ions in experiments on clean metals. This does indeed increase the ion yields, but the interpretation of the results is correspondingly complicated.

Several variations of the SIMS concept have been devised. In secondary ion imaging mass spectrometry (SIIMS), the mass filter is tuned to a particular element, and the relatively large area covered by the incident ion beam is imaged on a fluorescent screen by the ion optics (Castaing and Slodzian, 1962). An image of the spatial distribution of a particular constituent is thus obtained. The spatial distribution can also be obtained with the ion microprobe mass analyzer (IMMA). In this case the ion beam is focused to a diameter as small as $2 \mu m$, and the beam scans the surface in a rastor of perhaps $100 \mu m$ on an edge. Because the beam is scanned over a large area, the damage introduced by the concentrated beam is not great. Alternatively, the position of the beam can be fixed to obtain a profile into the sample. The ion microprobe mass analyzer has been reviewed by Anderson and Hinthorne (1972).

A spectrum of $GdCo_5$ is shown in Fig. 13. The spectrum was taken by

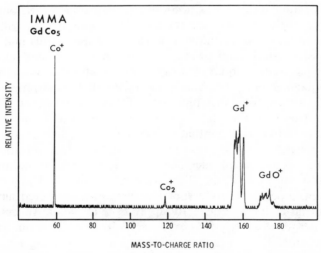

Fig. 13. Secondary ion mass spectrum of a $GdCo_5$ surface obtained by J. W. Guthrie, Sandia Laboratories, with the ion microprobe mass analzer. The spectrum was obtained with a $3.2 \, nA \, ^+N_2$ beam after the surface had been cleaned by sputtering. All five principal Gd isotopes—155, 156, 157, 158, and 160—are at least partially resolved. In addition to single ions of ^{59}Co there is a peak at Co_2^+. The group of peaks above the Gd^+ spectrum is due to GdO^+, suggesting that the oxygen observed in all of the $GdCo_5$ spectra is associated primarily with Gd atoms.

J. W. Guthrie at Sandia Laboratories using the ion microprobe (IMMA),
but it is typical of spectra taken by SIMS. All five principal isotopes of
Gd—155, 156, 157, 158, and 160—are at least partially resolved. In addition
to single ions of ^{59}Co, there is a peak due to Co_2^+. The group of peaks above
the Gd^+ spectrum is due to GdO^+. It is interesting that all of the surface
analytical techniques found oxygen apparently distributed throughout the
bulk of the $GdCo_5$ sample. Thus a residue of oxygen was detectable even
after sputter cleaning. From the IMMA spectrum it appears that the oxygen
is associated entirely with the Gd, since no cobalt–oxygen clusters were
observed. This kind of chemical information is not easily come by with other
techniques.

What appears to be needed at this point is an efficient means of ionizing
sputtered neutrals in the mass analyzer so that the ion yield can be directly
compared to the total yield.

2. The Atom-Probe Field Ion Microscope and Field Desorption Spectrometry

The ultimate in microanalytical sensitivity is the identification of single
atoms. Single-atom imaging has, of course, been available since the invention
by Müller (1956) of the field ion microscope. "That such imaging should
first be achieved, not with computer-designed electron lenses presided over
by teams of skilled technicians, but rather with an instrument of unparalleled
simplicity conceived and built by a single individual in a modest university
laboratory, stands as a classic example of intellect triumphing over the
bureaucratic system of financial support" (Duke and Park, 1972). Perhaps
it should not be surprising that single-atom identification should emerge
from the same fertile source, or that it should be accomplished with the same
elegant simplicity.

The atom-probe field ion microscope (Müller *et al.*, 1968) combines the
field ion microscope with a time-of-flight mass spectrometer. By manipulating
the specimen, a selected image spot is constrained to fall within a small
aperture in the imaging system. By applying a high voltage pulse to the
specimen, a layer of ions is field-evaporated. To the extent that the field-
evaporated ions follow the same trajectories as the imaging ions, the ion
corresponding to the preselected image spot will pass through the aperture
and enter the drift tube of a time-of-flight mass spectrometer. The drift tube
is conventionally about 2 m in length. By timing the interval between the
evaporation pulse and the time of arrival of the ion at a detector, the
mass-to-charge ratio of the ion can be computed from its known energy.

The mechanism by which a lattice is disassembled by the application of a high field is not fully understood. The two models that have been proposed, the image potential and the charge exchange models, have been subjected to detailed comparison with experiment by Vesely and Ehrlich (1973). They conclude that the image potential model, in which the limiting step is the escape of an ion over a Shottky barrier resulting from the superimposition of the potential created by the applied field and the classical image potential, provides the most satisfactory representation of their data on the field dependence of evaporation.

Complications are introduced by the phenomena of field adsorption of the noble image gas (Müller *et al.*, 1969) and apparent energy deficits of evaporated ions. The latter effect may be attributable to the excitation of plasmons by the escaping ion, as Lucas (1971) proposed for field-ionized gas atoms, but Müller (1971) suggests that a more likely explanation is premature field evaporation during the rise of the desorption pulse. Whatever the cause of the energy deficit, the uncertainty it introduces makes unit mass resolution very questionable. In fact, unit mass resolution is generally unnecessary except when dealing with the metal hydrides.

The principal factor limiting the application of the conventional atom-probe to surface problems is that from an image of perhaps 10^4 spots representing a given surface condition, only a single image spot or small group of spots can be selected for analysis. The surface is then completely destroyed by the act of analysis. The surface condition must therefore be reestablished and the experiment repeated a statistically significant number of times in order to arrive at a meaningful physical interpretation. The results of atom-probe investigations are thus generally presented in the form of histograms of the abundance of evaporated ionic species (Müller, 1971) and elaborate automation procedures are devised to repetitively perform the measurements (Turner *et al.*, 1973).

This limitation has been overcome by one of Müller's former students with the development of field desorption spectrometry (Panitz, 1974). This technique permits the complete crystallographic dependence of a given field-evaporated species to be obtained during a single evaporation event. This is accomplished in a modified field-ion microscope without the necessity of a separate drift tube by incorporating several technological innovations.

The microscope uses a unique, spherically curved Chevron channel electron multiplier array (CEMA). Two concentric CEMA plates intensify the image with a gain of about 4×10^6. This permits the microscope to be operated at helium image gas pressures of only 10^{-7} Torr. The fluorescent screen is also spherically curved and uses fiber-optic coupling to a flat viewing surface. A helium ion image of a tungsten surface at $27°$ K, obtained by J. A. Pantiz at Sandia Laboratories using this type of imaging, is shown in Fig. 14a.

(a)   (b)   (c)

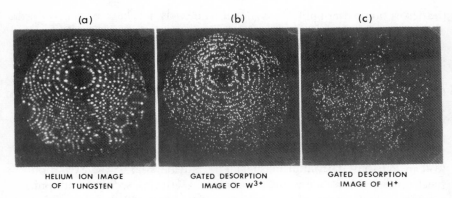

HELIUM ION IMAGE   GATED DESORPTION   GATED DESORPTION
OF TUNGSTEN   IMAGE OF W^{3+} IMAGE OF H^+

Fig. 14. Field-ion and field-desorption micrographs of a tungsten tip obtained by J. A. Panitz of Sandia Laboratories with the field desorption spectrometer. The helium ion image was obtained with the tip at $27°$ K at a helium pressure of only 10^{-7} Torr. The low image gas pressure is possible because of the 4×10^6 gain afforded by two spherically curved channel electron multiplier arrays. The gated desorption images of W^{3+} and H^+ were obtained with the spectrometer evacuated. The disorder of the W^{3+} desorption image relative to the helium-ion image suggests that the identification of individual field-ion image spots may be more difficult than is generally supposed. The gated desorption image of hydrogen exhibits even less order, indicating that hydrogen is randomly distributed over the surface of the tip. The principal hydrogen species is probably H^+, but the spectrometer is not capable of resolving H^+ from H_2^+, for example.

Several modes of operation as a field desorption spectrometer are possible. Unlike the conventional atom-probe, the desorption pulse is applied to an electrode very near the specimen, rather than to the specimen itself. The desorbed ions pass through an aperture in the electrode and are then decelerated and allowed to drift to the detector at a lower energy. This avoids the pulse calibration problems of previous atom-probes (Panitz *et al.*, 1969) and gives satisfactory resolution with only a 10 cm drift space (Panitz, 1973). The specimen is at the center of curvature of the CEMA so that travel times to any point of the detectors are equal.

The output of the multiplier array thus provides a time-of-flight record of all the evaporated ions reaching the detector. In addition, the ions produce visible scintillations on the screen. These can be detected with an external photomultiplier whose output is differentiated to record only the fast rise of the scintillation. Ionic species corresponding to several crystallographic areas can be identified simultaneously by using more than one photomultiplier (Panitz, 1973).

In an elegant extension of this technique the second CEMA plate is time gated to coincide with a preselected ionic species (Panitz, 1974). One thus obtains the complete desorption pattern of a given ionic species. In Fig. 14b the desorption pattern of W^{3+} is shown. The noticeable lack of a one-to-one correspondence between gated-desorption and ion-micrograph images suggests that using the atom-probe to identify individual ion-micrograph image spots may be more difficult than originally anticipated. Figure 14c shows the image corresponding to the desorption of hydrogen from the tungsten tip. The pattern exhibits no symmetry at all, from which it is concluded that hydrogen adsorbs at random interstitial positions. Previous desorption microscopes (Walko and Müller, 1972) were not time gated and thus imaged all desorbed species.

As fundamental understanding of the field-evaporation process increases, we can expect the atom-probe field ion microscope and field desorption spectrometry to provide us with quantitative measures of atomic binding energies.

IV. COMPARISONS

John Fortesque wrote in 1471 that "comparisons are odious." The passage of five centuries has not made them less so. I am therefore reluctant to add to the simplistic scorecards that usually climax reviews of analytical techniques. I was once asked by the author of such a review if I would provide him with the sensitivity of a particular instrument in "volts per second." Comparisons of that sort are not only odious, they are meaningless. It is important, therefore, to define some of the parameters that enter into a comparison.

A more difficult task is to distinguish between techniques and instruments. To what extent are the experimental limitations we observe intrinsic characteristics of the technique? For example, consider the uncertainty over the energy deficit observed in the operation of the atom probe (Section III.B.2). If it is due to the rise time of the desorption pulse, it can in principle be made arbitrarily small by better design. If it is due to plasmon excitation, however, it places a limit on the attainable mass resolution. Much work remains to be done before the intrinsic limitations of any of our analytical techniques are fully known, and comparisons made in the meantime are subject to drastic revision.

Comparisons are further complicated by the transmutability of instrument parameters. For example, the experimentalist is usually able to barter sensitivity for speed (bandwidth) over a very wide range. The relative importance of these parameters depends sensitively on the experiment being performed.

A. Sensitivity

The sensitivity of a surface analytical technique is properly stated in terms of the smallest percent of a monolayer on which a meaningful measurement can be made. It is limited by the background on which the signal appears. Background consists of unwanted information that is correlated with the desired signal, such as the bremsstrahlung background above which the characteristic x-ray excitation edges must be detected in SXAPS (Section II.A.3). It is important to make the distinction between background and random noise. Uncorrelated noise can be reduced to any desired level by integrating over a sufficiently long period.

Sensitivity can be increased only by suppressing the background. In several of the techniques, IS (Section II.A.2), SXAPS (Section II.A.3), EEAES (Section II.B.1), background suppression is accomplished by selectively amplifying the higher frequency components of the spectrum, since the background is a relatively slowly varying function of energy as compared to the spectral features of interest. This selective amplification is achieved by differentiating, which weights each Fourier component of the spectrum by its frequency. If the frequency components of the background are comparable to those of the signal, as in the case of diffraction features in the low-energy portion of the disappearance potential spectrum (Section II.A.4), differentiation is futile.

If we exclude the atom-probe (Section III.B.2), which clearly wins any sensitivity contest, the variation in sensitivity of a technique for different adsorbate-substrate combinations is so great that comparisons in sensitivity between the techniques are of little value. That is to say, the choice of which analytical technique should be used depends entirely on what system is to be studied. If, for example, small traces of alkali metal compounds need to be detected, the best choice would almost certainly be to use SIMS (Section III.B.1) because of the high ionization probability of sputtered alkali metal atoms. If the problem involves the detection of noble metals, it would not be wise to use SXAPS (Section II.A.3).

Generally, however, there will be more than one technique with sufficient sensitivity for a given problem. The selection process must then include other factors.

B. Resolution

Resolution is a measure of our ability to restrict the range over which we detect correlations. Nature provides us with very clear goals for the resolving power of our techniques. For those techniques that depend on a determination of atomic mass, for example, there is no point in distinguishing mass differences less than a single atomic mass unit.

Unit mass resolution is clearly achievable in SIMS (Section III.B.1). In the atom-probe (Section III.B.2) it is less clear. If the field-evaporated ions couple strongly to the excitation of plasmons, it may limit the achievable resolution to about 2 amu. It is doubtful whether unit mass resolution can be attained in the binary scattering technique (Section III.A.1) except at very low mass numbers, due to the constraint of the target atom by its neighbors. It does not appear, however, that low-temperature measurements of resolution have yet been attempted. A final judgment on the resolution of the binary scattering must therefore await further experimentation.

In the core level spectroscopies the smallest energy difference that can be detected need be no less than the uncertainty imposed by the lifetime of a core hole. For the narrowest core levels this is about 0.2 eV. This kind of resolution can apparently be achieved by all of the core level spectroscopies that do not use solid state energy analysis. The resolution of XPS (Section II.A.1) and IS (Section II.B.2) is limited by the quadratic sum of the energy spread of the excitation source and the pass band of the spectrometer. In SXAPS (Section II.A.3) only the energy spread of the primary electrons is involved. In EEAES only the pass band of the spectrometer contributes. These factors can be made less than 0.2 eV, but at a price.

Elemental identification, however, does not generally impose very severe requirements on resolution. It is the secondary features related to chemical bonding that place stringent demands on resolution. For the purpose of elemental identification, the objective of high resolution is simply to reduce overlap between the spectra of different elements. The severity of the overlap problem is in direct proportion to the complexity of the spectrum. The x-ray photoelectron spectrum (Section II.A.1) is the most complex of the core level spectra, since it contains both the core level excitation peaks and the Auger electron spectrum (Fig. 2). The photoelectron peaks can usually be distinguished by their relative sharpness, however. The least complex spectra are provided by IS (Section II.A.2) and SXAPS (Section II.A.3), since only excitation features are present (Figs. 4 and 6).

In general, improvements in resolution tend to decrease the background contribution and hence increase sensitivity. A heavy price is paid nevertheless.

C. Bandwidth

Improvements in sensitivity and resolution are purchased with bandwidth. The bandwidth is essentially the reciprocal of the measurement time constant and is limited by the level of noise that can be tolerated. The ultimate bandwidth limitation is therefore determined by the statistical noise associated with the measurement of discrete particles (shot noise). In this limit we can define the bandwidth ω as

$$\omega = 1/T = \sigma I/(S/N)^2, \qquad (21)$$

where the cross section σ is the number of signal counts per probe event, I is the number of probing events per unit time (primary current), and S/N is the signal-to-shot-noise ratio. T is the measurement time.

It is inevitable that the atom-probe that represents the acme of sensitivity should also represent the nadir of bandwidth. It may be instructive to consider the problem of determining the elemental composition of an unknown surface with the atom-probe (Section III.B.2). In the act of identifying an atom, the surface is destroyed. Even if the surface condition could be restored time after time, it would still be necessary to repeat the measurement a statistically significant number of times. The signal-to-shot-noise ratio for a given constituent will go as the square root of the number of times it is counted (van der Ziel, 1954). Thus in order to determine with 90% accuracy that a particular constituent comprises 10% of the surface atoms, the single-atom measurement would have to be repeated 10^3 times. The usefulness of the atom-probe lies in the fact that, unlike the other techniques, it is not necessary to sample at random. One can select certain crystallographic locations for analysis. Thus the trade-off between sensitivity and bandwidth determines what sort of problems a technique is best suited to handle. Discounting the patience of the scientist, the minimum bandwidth that can be tolerated in an experiment is fixed by the contamination rate of the surface.

The bandwidth defined by Eq. (21) is directly proportional to the rate at which the surface is probed. Thus by doubling the flux of primary particles—electrons, photons, or ions—a given signal-to-noise ratio is reached in half the time. There are, however, practical limitations on the flux of primary particles. We cannot, for example, increase the current in an electron or ion beam indefinitely without increasing the beam diameter because of space charge. This in turn degrades the resolution in IS (Section II.A.2) and EEAES (Section II.B.1). A more serious problem may be the damage inflicted by the incident particles.

D. Effect of the Probe

Quantum mechanics is a response to the realization that the properties of a system are inseparable from the act of observation. The interaction between our probes and the surfaces we are trying to understand is the most vexing aspect of surface analysis. Bizarre metal–noble gas molecular ions are created, for example, by the high fields required for the atom-probe field ion microscope. Although such effects in themselves are interesting, they may seriously confuse the study of phenomena of more direct interest.

The damage inflicted by the probe is a fairly obvious problem in SIMS and in the atom-probe where destruction of the surface is a deliberate act. But destruction may be no less complete in those techniques where it is unintentional. Ion bombardment causes sputtering, and electron impact can induce desorption. The latter effect imposes severe limitations on the use of IS, SXAPS, and EEAES to study chemisorption.

If the damage per incident particle is constant, the damage per unit area inflicted during a measurement can only be lessened by increasing the area probed. In those techniques that use deflection analyzers (XPS, IS, AES, ISS), this means a sacrifice of resolution. If the area covered by the incident beam is held constant and the current reduced, the time required to achieve a given signal-to-noise ratio increases such that the damage per unit area is undiminished. Techniques that do not use momentum dispersive analyzers (SXAPS, PIX) can tolerate an increase in the area covered by the primary beam without penalty. Thus the primary current density in appearance potential spectroscopy may be less than in EEAES, even though the current is two orders of magnitude greater. It is also possible to use a broad incident beam in the static method of SIMS because of the comparative insensitivity of the quadrupole mass analyzer to the initial momentum of an ion.

V. SUMMARY

Faced with a problem involving the chemical analysis of a solid surface, the scientist has a choice between some dozen possible analytical techniques. No simple rules can be stated to assist him in that choice. Each technique provides somewhat different information, and each is hampered by its own unique uncertainties. The scientist should also have the freedom to choose the optimum trade-off between instrument parameters to match his particular problem. This choice, however, is too often made for him by the equipment manufacturers.

Much of the discussion in this chapter dealt with the secondary features of the spectra that contain information on the chemical environment of the atoms. These effects must be viewed with some ambivalence, since they frequently interfere with quantitative determinations of elemental abundance and are generally imperfectly understood. This will be corrected only by comparative studies aimed at reconciling the observations made by different techniques and performed by scientists who construct their own instruments.

ACKNOWLEDGMENTS

The author is indebted to L. H. Scharpen of the Hewlett-Packard Company, and to W. Bauer, J. W. Guthrie, J. E. Houston, M. L. Knotek, R. G. Musket, G. C. Nelson, J. A. Panitz, and D. G. Schreiner of Sandia Laboratories, for providing unpublished spectra obtained by the various techniques, and for their comments on the manuscript.

REFERENCES

ANDERSON, C. A. (1970). *Int. J. Mass Spectrom. Ion Phys.* **3**, 413.
ANDERSON, C. A., and HINTHORNE, J. R. (1972). *Science* **175**, 853.
AMELIO, G. F. (1970). *Surface Sci.* **22**, 301.
AMELIO, G. F., and SCHEIBNER, E. J. (1968). *Surface Sci.* **11**, 242.
ARMOUR, D. G., and CARTER, G. (1972). *J. Phys. E: Sci. Instrum.* **5**, 2.
ASAAD, W. N. (1970). *Colloq. Int. Cent. Nat. Rech. Sci., Proc. Electron. Simples Multiples Domaine X X-UV, Paris 1970.*
BAER, Y., HEDÉN, P. F., HEDMAN, J., KLASSON, M., NORDLING, C., and SIEGBAHN, K., (1970). *Solid State Commun.* **8**, 517.
BAUER, W., and MUSKET, R. G. (1973). *J. Appl. Phys.* **44**, 2606.
BEARDEN, J. A., and BURR, A. F. (1967). *Rev. Mod. Phys.* **39**, 125.
BENNINGHOVEN, A. (1970). *Z. Phys.* **230**, 403.
BENNINGHOVEN, A. (1973). *Surface Sci.* **35**, 427.
BERGSTRÖM, I., and HILL, R. D. (1954). *Ark. Fys.* **8**, 21.
BISHOP, H. E., and RIVIÉRE, J. C. (1970). *Appl. Phys. Lett.* **16**, 21.
BRONGERSMA, H. H., and MUL, P. M. (1972). *Chem. Phys. Lett.* **14**, 380.
BRONGERSMA, H. H., and MUL, P. M. (1973). *Surface Sci.* **35**, 393.
BRUNDLE, C. R., and ROBERTS, M. W. (1972). *Proc. Roy. Soc. London A* **331**, 383.
BRUNDLE, C. R., and ROBERTS, M. W. (1973). *Chem. Phys. Lett.* **18**, 380.
BURHOP, E. H. S., and ASAAD, W. N. (1972). *Advan. in At. Mol. Phys.* **8**.
CARTER, G. (1973). *J. Vac. Sci. Technol.* **10**, 95.
CASTAING, R., and SLODZIAN, G. (1962). *J. Microsc.* **1**, 395.
CHANG, J., and LANGRETH, D. C. (1972). *Phys. Rev. B* **5**, 3512.
CHUNG, M. F., and JENKINS, L. H. (1970). *Surface Sci.* **22**, 479.
COAD, J. P., and RIVIÉRE, J. C. (1971). *Phys. Lett.* **35A**, 185.
CUTHILL, J. R., MCALISTER, A. J., and WILLIAMS, M. L. (1970). *J. Appl. Phys.* **39**, 1810.
DEV, B., and BRINKMAN, H. (1970). *Ned. Tijdschr. Vacuumtechn.* **8**, 176.
DONIACH, S., and ŠUNJIC, M. (1970). *J. Phys. C.* **3**, 285.
DUKE, C. B., and PARK, R. L. (1972). *Phys. Today* **25**, 23 August.
DUKE, C. B., and TUCKER, C. W. (1969). *Phys. Rev. Lett.* **23**, 1169.
ELLIS, W. P., and CAMPBELL, B. D. (1968). *J. Appl. Phys.* **41**, 1858.
FADLEY, C. S., and SHIRLEY, D. A. (1968). *Phys. Rev. Lett.* **21**, 980.
FAHLMAN, A., NORDBERG, R., NORDLING, C., and SIEGBAHN, K. (1966). *Z. Phys.* **192**, 476.
FRASER, W. A., FLORIO, J. V., DELGASS, W. N., and ROBERTSON, W. D. (1973). *Surface Sci.* **36**, 661.
GALLON, T. E. (1972). *J. Phys. D: Appl. Phys.* **5**, 822.
GERLACH, R. L. (1971). *Surface Sci.* **28**, 648.
GERLACH, R. L., and DUCHARME, A. R. (1972). *Surface Sci.* **32**, 329.
GERLACH, R. L., and TIPPING, D. W. (1971). *Rev. Sci. Instrum* **42**, 151.

GERLACH, R. L., HOUSTON, J. E., and PARK, R. L. (1970). *Appl. Phys. Lett.* **16**, 179.
GOFF, R. F. (1972). *J. Vac. Sci. Technol.* **9**, 154.
HAAS, T. W., and GRANT, J. T. (1970). *Appl. Phys. Lett.* **16**, 172.
HAGSTRUM, H. D. (1954). *Phys. Rev.* **96**, 336.
HAGSTRUM, H. D., and BECKER, G. E. (1971). *Phys. Rev. B* **4**, 4187.
HARRIS, L. A. (1968). *J. Appl. Phys.* **39**, 1419.
HAYDOCK, R., HEINE, V., and KELLY, M. J. (1972). *J. Phys. C* **5**, 2845.
HEILAND, W., SCHÄFFLER, H. G., and TAGLAUER, E. (1973). *Surface Sci.* **35**, 381.
HILL, R. D., CHURCH, E. L., and MIHELICH, J. W. (1952). *Rev. Sci. Instrum.* **23**, 523.
HONIG, R. E. (1958). *J. Appl. Phys.* **29**, 549.
HOUSTON, J. E. (1973). *Surface Sci.* **38**, 283.
HOUSTON, J. E., and PARK, R. L. (1969). *Appl. Phys. Lett.* **14**, 358.
HOUSTON, J. E., and PARK, R. L. (1970). *Bull. Amer. Phys. Soc.* **14**, 769.
HOUSTON, J. E., and PARK, R. L. (1971). *J. Vac. Sci. Technol.* **8**, 91.
HOUSTON, J. E., and PARK, R. L. (1972a). *Rev. Sci. Instum.* **43**, 1437.
HOUSTON, J. E., and PARK, R. L. (1972b). *Solid State Commun.* **10**, 91.
HOUSTON, J. E., and PARK, R. L. (1972c). *Phys. Rev. B* **5**, 3808.
HOUSTON, J. E., PARK, R. L., and LARAMORE, G. E. (1973). *Phys. Rev. Lett.* **30**, 846.
HÜFNER, S., COHEN, R. L., and WERTHEIM, G. K. (1972). *Physica Scripta* **5**, 91.
KIRSCHNER, J., and STAIB, P. (1973). *Phys. Lett.* **42A**, 335.
KLASSON, M., HEDMAN, J., BERNDTSSON, A., NILSSON, R., NORDLING, C., and MELNIK, P. (1972). *Physica Scripta* **5**, 93.
KORBER, H., and MEHLHORN, W. (1966). *Z. Phys.* **191**, 217.
LANDER, J. J. (1953). *Phys. Rev.* **91**, 1382.
LANGRETH, D. C. (1971). *Phys. Rev. Lett.* **26**, 1229.
LARAMORE, G. E. (1972). *Solid State Commun.* **10**, 85.
LEDER, L. B., and SIMPSON, J. A. (1958). *Rev. Sci. Instrum.* **29**, 571.
LIEFELD, R. J. (1968). *In* "Soft X-Ray Band Spectra" (D. J. Fabian, ed.). Academic Press, New York.
LUCAS, A. A. (1971). *Phys. Rev. Lett.* **26**, 813.
MCKAY, K. G. (1948). *Advan. Electron.* **1**, 72–83.
MADEY, T. E., YATES, J. T., and ERICKSON, N. E. (1973). *Chem. Phys. Lett.* **19**, 487.
MERZBACHER, E., and LEWIS, H. W. (1958). *In* "Encyclopedia of Physics" (S. Flügge, ed.), Vol. 34, p. 166. Springer-Verlag, Berlin.
MEYER, F., and VRAKKING, J. J. (1972). *Surface Sci.* **33**, 271.
MEYER, O., GYULAI, J., and MEYER, J. W. (1970) *Surface Sci.* **22**, 263.
MULARIE, W. M., and PERIA, W. T. (1971). *Surface Sci.* **26**, 125.
MÜLLER, E. W. (1956). *Phys. Rev.* **102**, 618.
MÜLLER, E. W. (1971). *Ber. Bunsengese. Phys. Chem.* **75**, 979.
MÜLLER, E. W., PANITZ, J. A., and MCLANE, S. B. (1968). *Rev. Sci. Instrum.* **39**, 83.
MÜLLER, E. W., MCLANE, S. B., and PANITZ, J. A. (1969). *Surface Sci.* **17**, 430.
MUSKET, R. G., and BAUER, W. (1972). *Appl. Phys. Lett.* **20**, 411.
NATTA, M., and JOYES, P. (1970). *J. Chem. Solids* **31**, 477.
NILSSON, P. O., and KANSKI, J. (1972). *Phys. Lett.* **41A**, 217.
NILSSON, P. O., and KANSKI, J. (1973). *Surface Sci.* **37**, 700.
PALMBERG, P. W. (1973). *Anal. Chem.* **45**, 549A.
PALMBERG, P. W., and RHODIN, T. N. (1968). *J. Appl. Phys.* **39**, 2425.
PANITZ, J. A. (1973). *Rev. Sci. Instrum.* **44**, 1034.
PANITZ, J. A. (1974). *J. Vac. Sci. Technol.* **11**, 206.
PANITZ, J. A., MCLANE, S. B., and MÜLLER, E. W. (1969). *Rev. Sci. Instrum.* **40**, 1321.

PARK, R. L., and HOUSTON, J. E. (1971). *Surface Sci.* **26**, 664.
PARK, R. L., and HOUSTON, J. E. (1972). *Phys. Rev. B* **6**, 1073.
PARK, R. L., and HOUSTON, J. E. (1973a). *J. Vac. Sci. Technol.* **10**, 176.
PARK, R. L., and HOUSTON, J. E. (1973b). *J. Appl. Phys.* **44**, 3810.
PARK, R. L., and HOUSTON, J. E. (1973c). *Phys. Rev. A* **7**, 1447.
PARK, R. L., HOUSTON, J. E., and SCHREINER, D. G. (1970). *Rev. Sci. Instrum.* **41**, 1810.
PARRATT, L. G. (1959). *Rev. Mod. Phys.* **31**, 616.
PICRAUX, S. T., and VOOK, F. L. (1971). *Appl. Phys. Lett.* **18**, 191.
POWELL, C. J. (1973). *Phys. Rev. Lett.* **30**, 1179.
POWERS, D., and WHALING, W. (1962). *Phys. Rev.* **126**, 61.
REDHEAD, P. A., and RICHARDSON, G. W. (1972). *J. Appl. Phys.* **43**, 2970.
RICHARDSON, O. W. (1928). *Proc. Roy. Soc. A* **119**, 531.
RICHARDSON, O. W., and BAZZONI (1921). *Phil. Mag.* **42**, 1015.
ROBINSON, H. R., and YOUNG, C. L. (1930). *Proc. Roy. Soc. A* **128**, 92.
SAR-EL, H. Z. (1967). *Rev. Sci. Instrum.* **38**, 1210.
SARIS, F. W. (1972). *In* "Physics of Electronic and Atomic Collisions," VII, ICPEAC 1971. North-Holland Publ., Amsterdam.
SCHROEER, J. M., RHODIN, T. N., and BRADLEY, R. C. (1973). *Surface Sci.* **34**, 571.
SEWELL, P. B., and COHEN, M. (1967). *Appl. Phys. Lett.* **11**, 298.
SHÖN, G., and LUNDIN, S. T. (1973). *J. Electron Spectrosc.* **1**, 105.
SICKAFUS, E. N. (1973). *Phys. Rev. B* **7**, 5100.
SIEGBAHN, K., NORDLING, C., FAHLMAN, A., NORDBERG, R., HAMRIN, K., HEDMAN, J., JOHANSSON, G., BERGMARK, T., KARLSSON, E., LINDGREN, I., and LINDBERG, B. (1967). *In* "ESCA: Atomic, Molecular, and Solid State Structure by Means of Electron Spectroscopy." Almqvist and Wiksells, Uppsala.
SIEGBAHN, K., HAMMOND, D., FELNER-FELDEGG, H., and BARNETT, E. F. (1972). *Science* **176**, 245.
SIEGBAHN, M. (1931). *In* "Spektroscopie der Röntgenstrahlung." Springer Verlag, Berlin.
SMITH, D. P. (1967). *J. Appl. Phys.* **38**, 340.
SMITH, D. P. (1971). *Surface Sci.* **25**, 171.
STEINHARDT, R. G., and SERFASS, E. H. (1951). *Anal. Chem.* **23**, 1585.
STEINHARDT, R. G., HUDIS, J., and PERLMAN, M. J. (1972). *Phys. Rev. B* **5**, 1016.
SZALKOWSKI, F. J., and SOMORJAI, G. A. (1972). *J. Chem. Phys.* **56**, 6097.
TARNG, M. L., and WEHNER, G. K. (1973). *J. Appl. Phys.* **44**, 1534.
TAYLOR, N. J. (1969). *J. Vac. Sci. Technol.* **6**, 241.
TRACY, J. C. (1971). *Appl. Phys. Lett.* **19**, 353.
TRACY, J. C. (1972). *J. Appl. Phys.* **43**, 4164.
TRACY, J. C., and PALMBERG, P. W. (1969). *J. Chem. Phys.* **51**, 4582.
TURNER, P. J., REGAN, B. J., and SOUTHON, M. J. (1973). *Surface Sci.* **35**, 336.
VAN DER WEG, W. F., KOOL, W. H., and ROOSENDAAL, H. E. (1973). *Surface Sci.* **35**, 413.
VAN WIJNGAARDEN, A., MIREMADI, B., and BAYLIS, W. E. (1971). *Can. J. Phys.* **49**, 2440.
VESELY, M., and EHRLICH, G. (1973). *Surface Sci.* **34**, 547.
WAGNER, C. D., and BILOEN, P. (1973). *Surface Sci.* **35**, 82.
WALKO, R. J., and MÜLLER, E. W. (1972). *Phys. Status Solidi.* (a) **9**, K9.
WEBER, R. E., and PERIA, W. T. (1967). *J. Appl. Phys.* **38**, 4355.
WERNER, H. W. (1969). *Develop. Appl. Spectrosc.* **7A**, 252.
WERNER, H. W., and DE GREFTE, H. A. M. (1973). *Surface Sci.* **35**, 458.

9

Surface Vibrations

M. G. LAGALLY[†]

COLLEGE OF ENGINEERING
UNIVERSITY OF WISCONSIN
MADISON, WISCONSIN

I. INTRODUCTION

In nearly all aspects of the properties of, reactions at, or interactions with a crystalline surface, the temperature is an important parameter. Surface thermal properties are considerably different from those of the bulk. Because the surface is a defect (with changed force constants and possibly different

[†] Alfred P. Sloan Foundation fellow.

geometry) in an otherwise perfect infinite crystal, local modes of vibration are present with frequencies outside the range of the bulk modes, mean square vibrational amplitudes of surface atoms are larger than in the bulk, and anharmonicity becomes important at much lower temperatures. As a result, surface vibrations are important in a variety of ways. For example, transport properties at surfaces are affected by the thermal vibrations of surface atoms, as is the scattering of atoms and low-energy electrons. The thermodynamic properties of very small particles and very thin films are in large part determined by the surface vibrations. The transition temperature in thin-film superconductors may be higher than in thick films because of the presence of surface modes. A whole technology in device physics rests on the excitation of surface waves in piezoelectric crystals. On a grandiose scale, surface waves with 400 km wavelength traveling along the earth's crust are of interest in seismology.

In this chapter we discuss the vibrational properties of clean crystalline surfaces. Contaminated surfaces, adsorbed layers, very thin films, or composite structures, although interesting and important in their own right, fall outside the scope of this chapter. Aspects of these topics are treated elsewhere in these two volumes. Within these limitations, we have attempted to select theoretical and experimental work that provides the most simple physical insight. In order to lay the groundwork for discussion of vibrational motion, in Section II we briefly treat the static crystal. Methods for the calculation of the normal modes of vibration for a crystal with surfaces are discussed in Section III. Theories, advanced to the stage where the mode spectrum can be calculated throughout the Brillouin zone, show that surface modes are likely to occur anywhere in gaps in the bulk phonon bands, and that the dispersion relations may be quite different from those of an infinite crystal. Because this changes the frequency distribution function, the thermodynamic properties of a crystal with surfaces must be different from an infinite crystal, as must the surface mean square vibrational amplitudes. These are discussed in Sections IV and V, respectively. The difficulty in obtaining a large and well-defined surface area–to–volume ratio makes the surface contribution to thermodynamic functions difficult to measure. This is not true for surface mean square vibrational amplitudes, because the surface can be effectively isolated from the bulk with the scattering of low-energy electrons, which penetrate only a few atomic layers. The scattering of slow electrons has been the single most successful technique for studying surface vibrations. Its use for measuring mean square vibrational amplitudes and details of the phonon spectrum is discussed in Sections VI and VII. Finally, because of its promise as a tool for learning about temperature effects on interactions with surfaces, atomic-beam scattering is discussed briefly in Section VIII.

Since a review such as this can never be exhaustive and always reflects the author's own bias and greater familiarity with some parts of the subject, a list of related reviews is given at the end of the references. The reader may wish to turn to these for a more extensive treatment of particular aspects of this topic.

II. THE STATIC–CRYSTAL SURFACE

Before considering thermal vibrations of surfaces, we briefly review some aspects of the structure and thermodynamics for static surfaces. These are discussed in more detail elsewhere (Strozier *et al.* and Leamy *et al.*, Volume I) and have been reviewed earlier (Benson and Yun, 1967). The creation of a surface requires the removal of atoms on one side of a plane. In order to achieve mechanical equilibrium with the lowered symmetry for atoms near the surface, the force constants may change and the surface atoms may relax from their bulk position. Such force–constant changes and atom displacements of course affect the surface phonon spectrum and hence the vibrational properties of the surface.

The minimum-energy configuration for a rigid crystal with a surface requires a surface contribution $\Phi^s(0)$ to the total energy. $\Phi^s(0)$ may be written in two parts

$$\Phi^s(0) = \Phi_0^s(0) + \Delta\Phi^s(0), \tag{1}$$

where $\Phi_0^s(0)$ is the surface energy for the undistorted rigid crystal and $\Delta\Phi^s(0)$ is the correction due to the relaxation near the surface. To calculate this relaxation, it is usually assumed that each plane of atoms is displaced as a whole perpendicular to the surface. The total static energy of the crystal is then minimized with respect to the displacements δ_m from the bulk spacing (Alder *et al.*, 1959; Burton and Jura, 1967; Allen and de Wette, 1969a; Jackson, 1971; Cheng *et al.*, 1974). A typical behavior for the static displacements is shown in Fig. 1. Depending on the type of potential used and the surface orientation, the displacement of the outer plane is several percent of the bulk interlayer separation and falls rapidly with increasing distance from the surface. The displacement is in general positive, i.e., an expansion, except in ionic solids (Benson and Yun, 1967). Recently Cheng *et al.* (1974) have shown that for some surfaces of bcc metals, the sign of the displacement alternates with depth from the surface.

Surface energies calculated for static crystals vary over a wide range but roughly parallel the cohesive energies. Rare-gas crystals have low surface energies, followed by ionic and metallic crystals. The contribution to the

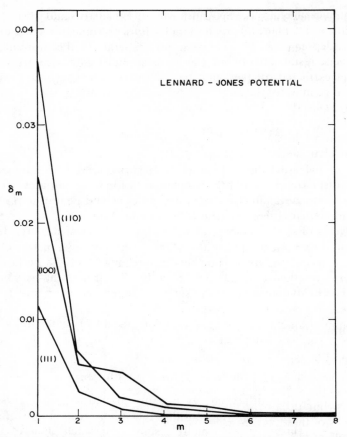

Fig. 1. Static displacements for the (100), (110), and (111) orientations of an fcc Lennard–Jones solid. δ_m is the fractional change from the bulk spacing and m labels the plane number. (After Allen and de Wette, 1969a.)

surface energy due to the relaxation $\Delta\Phi^s(0)$ is small for rare-gas crystals and metals, amounting to less than 1% and several percent respectively, but may be much larger, 20–40%, for ionic crystals.

Experimental estimates of the displacement of surface layers can in principle be obtained from low-energy electron diffraction (LEED) measurements, since a changed layer spacing will cause peak shifts and changes in peak shapes. This has been done either by comparing individual intensity versus energy profiles with dynamic theories of the scattering or by analysis of the kinematic or single-scattering part of the diffracted intensity.[†] Comparisons

[†] These methods, of course, give the displacement of surface layers at the temperature of the experiment. This is comparable to the static displacements only at low temperatures.

with dynamic calculations give values ranging from approximately -15% to 10% for the outer-plane displacement of several metals, with a sensitivity of about 5% (Jepsen *et al.*, 1972; Martin and Somorjai, 1973; Laramore, 1973). More precise results are obtained by analyzing the kinematic part of LEED intensities extracted by averaging the diffracted intensity over a range of diffraction geometries (Lagally *et al.*, 1971; Ngoc *et al.*, 1973). For example, Buchholz *et al.* (1975) obtain an upper limit on the outer-plane relaxation in $W(110)$ of 3%.

If the atoms in the crystal are now allowed to vibrate, a vibrational spectrum characteristic of the atomic potential near the surface results. Because of the anharmonicity of this potential, there will be a further, dynamic displacement, and mean square vibrational amplitudes of surface atoms will be much larger than the bulk. These cause temperature-dependent contributions to the surface thermodynamic functions and affect the scattering of slow electrons, atoms, and ions. We consider each of these subjects in turn.

III. SURFACE MODES OF VIBRATION

The surface of a crystal, since it destroys the perfect periodicity of the bulk, can be considered as a laterally extended defect. Therefore, just as point defects give rise to localized modes of vibration, there will be modes of vibration associated with the surface. These modes are wavelike in directions parallel to the surface, since the lattice periodicity is preserved in these directions, but decay rapidly with distance away from the surface. The frequencies of these modes generally lie below the range of allowed values for the bulk modes. This can be understood by recognizing that the creation of a surface requires the force constants across the dividing plane to go to zero.

Surface modes at long wavelengths can be treated by considering the crystal as an elastic continuum. However, for wavelengths shorter than $\sim 10^{-6}$ cm, lattice dispersion becomes important, the continuum treatment breaks down, and lattice-dynamical methods must be used. We turn first to a brief discussion of the continuum theory.

A. Elastic Continuum Theory of Normal Modes

The simplest method of calculating surface vibrational modes is to consider waves propagating in a semi-infinite isotropic medium that possesses a stress-free surface. The first investigation of this nature was by Lord

Rayleigh (1885), who discovered the long-wavelength acoustic modes known as Rayleigh waves. An isotropic elastic continuum was also considered by Stratton (1953), who pointed out that the introduction of a surface mixes waves of different polarization and propagation directions. The modes in the vicinity of a free surface can be classified into four types:

1. Surface modes, which are waves with propagation directions parallel to the surface and displacement amplitudes decaying exponentially with depth into the medium;

2. Transverse modes, which are bulk waves with polarization vectors parallel to the surface;

3. Mixed modes, which are a superposition of longitudinal surface and transverse bulk waves; and

4. Combination modes, which are a superposition of longitudinal bulk and transverse bulk waves.

Modes of the first type are the so-called Rayleigh waves. They have frequencies that tend to zero as $q_\parallel \to 0$ and hence are acoustic surface modes. Rayleigh waves are nondispersive and have velocities between 10^5 and 6×10^5 cm/sec.

Surface elastic waves in anisotropic media were investigated by Stoneley (1955) and later by others (Gazis *et al.*, 1960; Lim and Farnell, 1968, 1969; Burridge, 1970). If the medium is bounded by a stress-free surface at $z = 0$, the displacement $\mathbf{u}(\mathbf{r})$ at the point $\mathbf{r}$ in the medium is given by

$$\mathbf{u}(\mathbf{r}) = \sum_{n=1}^{3} a_n \zeta^n \exp[i(\mathbf{q}^n \cdot \mathbf{r} - \omega t)]$$

$$= \sum_{n=1}^{3} a_n \zeta^n \exp[i(q_z^n z + q_x x + q_y y - \omega t)], \tag{2}$$

where $(q_x, q_y) = \mathbf{q}_\parallel$ is the propagation vector parallel to the surface and q_z^n may be complex, ζ^n is the polarization vector, and a_n is a constant. For an isotropic medium, q_z^n's are pure imaginary, and the amplitudes decay exponentially with distance from the surface. These are the Rayleigh waves. If the q_z^n's are complex, the wave decays in an oscillatory manner with depth. The resulting waves are called generalized Rayleigh waves. If the q_z^n's are real, the associated displacements do not decay with distance from the surface and no surface wave exists. Figure 2 shows an example of the relationship between elastic constants and the type of wave excited for several materials.

It should be noted that q_z is always of the order of $\mathbf{q}$, the propagation vector, and thus the penetration is of the order of a wavelength. For wavelengths large compared to a lattice constant, the limit for which the continuum theory is valid, the penetration of the wave is deep. Hence the continuum model gives for the large part information about the displacements

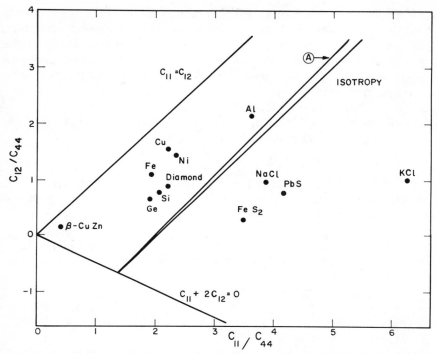

Fig. 2. Propagation of Rayleigh waves in the [100] direction in a number of materials. The region of crystal stability is to the right of the lines $c_{11} = c_{12}$ and $c_{11} + 2c_{12} = 0$. Generalized Rayleigh waves occur to the left of the line labeled A, and ordinary Rayleigh waves to the right. The line for isotropic materials lies entirely in the ordinary Rayleigh wave region. (After Gazis *et al.*, 1960.)

of atoms in the bulk, with bulk elastic constants determining the frequency of the mode. However, such long-wavelength modes are of great interest in seismology, ultrasonics, and device physics, and their properties have been studied extensively, both theoretically and experimentally [a list of references is given by Wolfram *et al.* (1972)]. In order to be sensitive to the surface proper, surface modes must have wavelengths comparable to a lattice constant. Because the crystal can then no longer be considered as a continuum and the phase velocity of a mode becomes a function of the wavelength, lattice-dynamical methods must be used to calculate the mode spectrum.

B. Lattice Dynamics

The theory of lattice dynamics deals with the equations of motion of atoms comprising the crystal. To obtain the spectrum of vibrational modes, one needs to specify the positions of the atoms and the forces acting between them. Defining the potential energy of the crystal in terms of the instantaneous positions of all the atoms, and assuming small vibrations, it is possible to expand the potential energy in a Taylor series about the equilibrium positions of the atoms. Neglecting all but the quadratic term then gives the equations of motion for the atoms in the harmonic approximation.

The calculation for a crystal with surfaces proceeds in the same manner (Maradudin *et al.*, 1971). The position of a particle vibrating about its equilibrium position $\mathbf{r}_0$ in the lattice is given by

$$\mathbf{r}(l, K) = \mathbf{r}_0(l, K) + \mathbf{u}(l, K), \tag{3}$$

where $\mathbf{u}(l, K)$ is the displacement from equilibrium and l, K specify the Kth particle in the lth unit cell. Dividing $\mathbf{r}_0(l, K)$ into components parallel and perpendicular to the surface gives

$$\mathbf{r}(l, K) = r_\perp(l_3, K) + \mathbf{r}_\parallel(l_\parallel, K) + \mathbf{u}(l, K). \tag{4}$$

The total kinetic energy of the lattice is given by

$$T = \tfrac{1}{2} \sum_{l, K, \alpha} M_K \dot{u}_\alpha^2(l, K), \tag{5}$$

where M_K is the mass of the Kth kind of particle and $\alpha = 1, 2, 3$ labels the cartesian component of $\mathbf{u}$.

The total potential energy Φ of the system is assumed (adiabatic approximation) to be a function of the instantaneous positions of the atoms. Φ is expanded in a Taylor series about the equilibrium positions

$$\Phi = \Phi_0 + \sum_{l, K, \alpha} \Phi_\alpha(l, K) u_\alpha(l, K)$$

$$+ \tfrac{1}{2} \sum_{\substack{l, K, \alpha \\ l', K', \beta}} \Phi_{\alpha\beta}(l, K; l', K') u_\alpha(l, K) u_\beta(l', K'), \tag{6}$$

where Φ_0 is the potential energy of the equilibrium configuration,

$$\Phi_\alpha(l, K) = \left(\frac{\partial \Phi}{\partial u_\alpha(l, K)} \right)_0, \tag{7}$$

and

$$\Phi_{\alpha\beta}(l, K; l', K') = \left(\frac{\partial^2 \Phi}{\partial u_\alpha(l, K) \partial u_\beta(l', K')} \right)_0. \tag{8}$$

The derivatives are evaluated at the equilibrium positions. Higher-order terms in Eq. (6) are assumed to be small and are neglected. $\Phi_\alpha(l, K)$, the component of the force in the α direction, vanishes for the equilibrium configuration. Hence the equations of motion are

$$M_K \frac{d^2}{dt^2}(u_\alpha(l, K)) = -\frac{\partial \Phi}{\partial u_\alpha(l, K)}$$

$$= -\sum_{l', K', \beta} \Phi_{\alpha\beta}(l, K; l', K') u_\beta(l', K'). \tag{9}$$

The coefficients $\Phi_{\alpha\beta}$ are the force constants. $-\Phi_{\alpha\beta}(l, K; l', K')$ is approximately the force exerted in the α direction on atom (l, K) when atom (l', K') is displaced a unit distance in the β direction while all other atoms remain fixed. In the harmonic approximation, the $\Phi_{\alpha\beta}$ are evaluated at the static equilibrium positions, which includes static relaxation of atoms near the surface but does not include thermal expansion. In the quasiharmonic approximation the $\Phi_{\alpha\beta}$ are evaluated at the "dynamic equilibrium" positions, which includes the bulk thermal expansion and any excess surface thermal expansion.

Solutions to Eq. (9) must satisfy the Bloch condition and are therefore of the form

$$u_\alpha(l, K) = \frac{u_\alpha(K)}{M_K^{1/2}} \exp[i(\mathbf{q} \cdot \mathbf{r}(l) - \omega t)], \tag{10}$$

where ω is the vibrational frequency and $u_\alpha(K)$ is the displacement of the Kth atom in the unit cell that represents the origin for the vectors $\mathbf{r}(l)$.

To take explicit account of the surface, $u_\alpha(l, K)$ is rewritten

$$u_\alpha(l, K) = u_\alpha(l_3, K) \exp[i(\mathbf{q} \cdot \mathbf{r}_\parallel(l) - \omega t)]. \tag{11}$$

where

$$u_\alpha(l_3, K) = (NM_K)^{-1/2} \xi_\alpha(l_3, K) \exp(iq_z r_\perp). \tag{12}$$

$\mathbf{q} \cdot \mathbf{r}_\parallel$ is the projection of the propagation vector onto the surface and q_z is again a real or complex attenuation factor describing the decay of the displacements with depth. As before if the q_z's are pure imaginary, the mode is a Rayleigh wave if $\omega \to 0$ as $\mathbf{q}_\parallel \to 0$ or a pure optical surface mode if $\omega \neq 0$. If q_z is complex, a generalized Rayleigh or optical surface wave results. ξ_α is the amplitude of vibration in the α direction of the mode at (l, K). Eqs. (9), (11), and (12) give as usual a set of equations

$$\sum D_{\alpha\beta} \xi_\beta = \omega^2 \xi_\alpha, \tag{13}$$

which have nontrivial solutions when the determinant of the coefficients vanishes. $D_{\alpha\beta}$, the elements of the dynamical matrix $\tilde{D}$, are given by

$$D_{\alpha\beta} = \frac{1}{M_K^{1/2}} \frac{1}{M_{K'}^{1/2}} \sum_{l'} \Phi_{\alpha\beta}(l, l') \exp[i\mathbf{q} \cdot (\mathbf{r}(l) - \mathbf{r}(l'))]. \tag{14}$$

The solutions of Eq. (13) determine the dispersion relations.

Gazis *et al.* (1960) have applied this formalism to semi-infinite monatomic model crystals to obtain the dispersion curve for the acoustic surface mode. Hooke's-law interactions of nearest and next-nearest neighbors in a simple cubic lattice were assumed. No account was taken of changes in geometry or force constants caused by lattice distortion near the surface. Physical parameters were chosen to fit data for KCl, which is nearly monatomic as far as the masses are concerned and which crystallizes in a simple cubic array. As expected, the surface mode approaches the continuum limit at small $|q_{||}|$ and begins to show dispersion as $|q_{||}|$ increases. Although dispersive, the mode is characteristic of a Rayleigh wave through about half the Brillouin zone and then becomes a generalized Rayleigh wave.

Wallis *et al.* (1968) have considered surface modes in a semi-infinite diatomic simple cubic crystal. Under the same constraints as above, the modes of vibration on the (100) face of NaCl were calculated. Because of complexities in the calculation for general $q_{||}$ for a semi-infinite crystal,

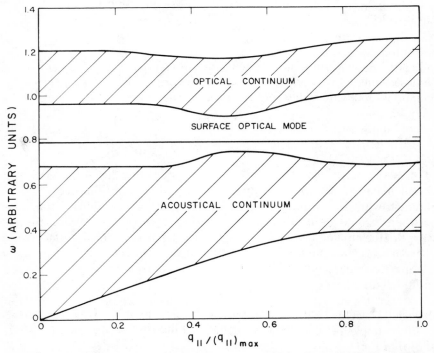

Fig. 3. Dispersion curve for optical surface waves propagating in the [100] direction on a (100) surface of an NaCl-type crystal with short-range forces only. Because for each $q_{||}$, a range of $q_{\perp}$ can contribute, bulk modes appear as bands with a wide frequency range. The optical surface mode lies below the bulk optical band. (After Wallis *et al.*, 1968.)

the calculation was carried out for an 18-layer slab with two free surfaces. Figure 3 shows the optical surface modes as well as the bulk optical and acoustical bands.

Allen *et al.* (1969, 1971a,b) have used similar slab-shaped crystals in extensive calculations of the normal-mode spectrum with more realistic model surfaces and atomic potentials. They point out the differences to be expected from using a slab instead of a semi-infinite crystal. (1) Since there are two free surfaces, there will be two nearly degenerate surface modes of each kind. In the limit of a very thick slab, these become degenerate. (2) Calculations for slabs are not valid for long wavelengths, since in this limit some surface modes have large penetration and will thus be affected by the finite thickness of the slab. (3) The finite number of layers in the slab limits the number of modes in the bulk bands, which depends on the number of possible $q_\perp$ for each $\mathbf{q}_\parallel$. This is infinite for a semi-infinite crystal; the effect of increasing the slab thickness is to populate the bulk bands more densely without, however, affecting the nonpenetrating (short-wavelength) surface modes. The limiting bulk band frequencies are also somewhat affected by the finite thickness of the slab. Within these restraints computer calculations with slab-shaped crystals appear to be useful for studies of vibrational properties of thick crystals, as well as for reconstructed surfaces and adsorbed layers, without a need for simplifying assumptions regarding crystal structure, interaction potential, or surface orientation. These calculations do point out that the vibrational properties of very thin films may be considerably different from those of thick crystals, because of the altered mode spectrum.

Allen *et al.* use the Lennard–Jones potential

$$\Phi(r) = 4\varepsilon[(\sigma/r)^{12} - (\sigma/r)^6] \tag{15}$$

to calculate the vibrational spectrum of a monatomic fcc slab. A particle interacts with all its neighbors, and displacements near the surface from bulk positions are taken into account. Because the results depend only on the shape of the potential and not on the potential parameters ε, σ or the mass of the particles, this potential is the simplest one suitable for obtaining a qualitative understanding of the surface properties of monatomic solids, although quantitatively the potential is applicable only to noble-gas solids, of course. Especially in metals, a model with central forces only is not expected to yield correct results.

An example of dispersion curves calculated with the above model is shown in Fig. 4. Figure 4a shows the dispersion relation for a 21-layer slab in the (111) orientation with, however, periodic boundary conditions in the z directions; i.e., surfaces are not allowed and no surface modes appear. In general three bulk bands can be identified, corresponding to a longitudinal

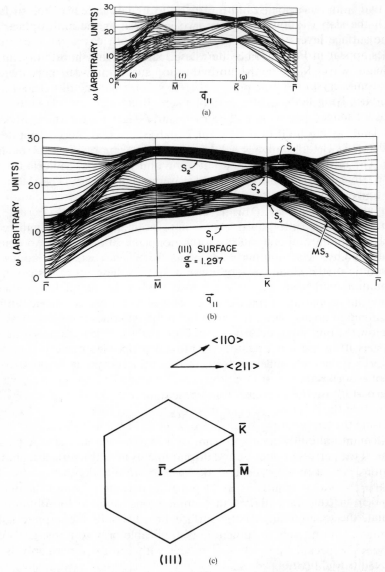

Fig. 4. Dispersion curve for modes in a 21-layer slab of a monatomic fcc Lennard–Jones solid in the (111) orientation. (a) Periodic boundary conditions in the z direction, bulk modes. Three bands can be identified, corresponding to longitudinal and transverse waves. (b) Mode spectrum with the surface taken into account. A variety of surface modes S appears. S_1 is the Rayleigh wave. (c) Schematic of the Brillouin zone. σ/a is the density of the crystal and corresponds here approximately to Xe at $0°$ K.(After Allen *et al.*, 1971b.)

and two transverse polarizations. This is to be compared with Fig. 4b, in which the slab is allowed to have two surfaces. The static displacement of the surface layers is taken into account in both cases. Several surface modes appear in Fig. 4b. The interesting feature is that in addition to the Rayleigh wave S_1, other, high-frequency surface modes can exist in monatomic crystals. These modes exist at short wavelengths near the zone boundary, lying between the bulk bands, and get absorbed into the bulk bands for longer wavelengths. They will therefore not arise in a continuum model.

Allen et al. (1971b) indicate a way to understand simply how these modes arise. The introduction of the surface represents a perturbation (1) because of the truncation of the crystal and (2) because of possible force–constant changes for atoms near the surface. When the modes penetrate deeply, the latter are not important; and the only effect is the setting to zero of the force constants across the interface. This gives rise to the Rayleigh mode, which is found for all models. The modes near the Brillouin zone boundary are affected by the second perturbation as well; its strength is related to the magnitude of the change in the force constants. For the Lennard–Jones potential, the creation of a surface allows surface atoms to relax outward, causing a weakening of the force constants and thus a lowering of the vibrational frequency at a given $\mathbf{q}_\parallel$. Hence these surface modes should also lie below the bulk bands. Thus, for example, in Fig. 4b, S_1 between $\bar{M}$ and $\bar{K}$ comes from the lower transverse bulk band, S_2 from the longitudinal band, etc. If the force constants were to become stronger (as possibly in the case of adsorption), the surface modes would appear above the bulk bands. If the perturbation representing the surface is weak, surface modes will be localized in the surface layer. As the perturbation is increased (for example, by raising the temperature to allow more dynamic displacement), additional surface modes should appear that are primarily localized in the next layer of atoms. If the perturbation is not strong enough to effectively localize a mode within a layer, a mixed bulk–surface mode results. An example is MS_3 in Fig. 4b.

To demonstrate the surface character of these modes, Allen et al. (1971b) calculate the squared amplitude of the mode in the mth layer,

$$|\xi(m)|^2 = |\xi_1(m)|^2 + |\xi_2(m)|^2 + |\xi_3(m)|^2. \tag{16}$$

The results for some of the surface modes of Fig. 4b plotted against layer number are shown in Fig. 5. The amplitude of the Rayleigh wave S_1 shows a nearly exponential decay into the bulk and, as expected, falls more slowly for large $\mathbf{q}_\parallel$. The other modes, although clearly localized in the surface layer, have a very complex decay with depth.

Changes in the surface force constants were studied by turning off the

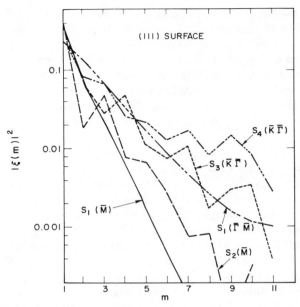

Fig. 5. Decay of the amplitude of the surface modes of Fig. 4b with depth into the crystal. m labels the plane. $\overline{\Gamma M}$ indicates **q** values midway between $\overline{\Gamma}$ and $\overline{M}$. (After Allen *et al.*, 1971b.)

static relaxation at the surface, making the interplanar distance the same throughout the crystal. Because this increases the force constants, the surface mode frequencies rise and in some cases become part of the bulk bands. This demonstrates the importance of considering also the dynamic displacements at the surface, as well as any changes in surface structure or binding. It is well known that many semiconductors and some metals have a rearranged surface structure. Further, the excess surface thermal expansion may be significant at temperatures where experiments are performed. The inclusion of these effects is likely to change considerably the calculated spectrum and may lead to the discovery of additional surface modes. Unfortunately, this must await a better experimental determination of the atomic structure and lattice dynamics of surfaces (Webb and Lagally, 1973).

Force constant changes have also been studied (Musser and Rieder, 1970) in models that include angular forces to simulate metals. Force constant values were chosen to match the elastic constants of Ni, and the central-force coupling between the outermost and second layers was weakened by an amount determined from low-energy electron diffraction measurements of the effective mean square vibrational amplitude of surface atoms (Section VI).

The dispersion curves were surprisingly similar to those of Allen *et al.* (1971b), suggesting that the Lennard–Jones potential may be qualitatively applicable to vibrational mode investigations in metals as well.

As pointed out earlier in this section, calculations for diatomic crystals are fundamentally no different, as long as only short-range interactions are considered. A different situation arises if long-range forces are taken into account. The next section discusses briefly the calculation of modes for this case.

C. Modes Depending on Long-Range Forces

In addition to the local atomic interactions, long-range Coulomb forces are frequently present, such as in ionic crystals, where the vibrations of the sublattices against each other produce an oscillating dipole moment in the unit cell. Even in monatomic crystals, where such forces are absent in the bulk, local distortions of the charge distribution near the surface (e.g., for a rearranged surface) may set up an electric field. Interaction with an electromagnetic field or a moving charged particle can therefore excite a new type of "surface" wave. These waves are essentially electrodynamic in origin and, because of this, are much easier to observe than the microscopic surfaces modes, at least in the long-wavelength limit (surface polaritons).

Calculations of the modes of vibration of ionic crystals have been made by both continuum (Fuchs and Kliewer, 1965; Kliewer and Fuchs, 1966a, b; Engleman and Ruppin, 1966, 1968a, b) and lattice–dynamical methods (Lucas, 1968; Tong and Maradudin, 1969; Jones and Fuchs, 1971; Chen *et al.*, 1970, 1971a, b, 1972a, b, c). Since the continuum model gives some simple insight into the new feature introduced by long-range forces, i.e., optically active surface phonons, we discuss it briefly. The normal modes of the coupled photon–phonon system can be obtained by solving Maxwell's equations for the electromagnetic field inside and outside the crystal. The equations of motion for an optically isotropic ionic crystal are (Born and Huang, 1954)

$$d^2\mathbf{w}/dt^2 = b_{11}\mathbf{w} + b_{12}\mathbf{w}, \qquad \mathbf{P} = b_{21}\mathbf{w} + b_{22}\mathbf{E}, \tag{17}$$

where $\mathbf{P}$ is the polarization and $\mathbf{w}$ is the relative displacement of the positive and negative sublattices, divided by the root of the reduced mass.

$$\mathbf{w} = (\mathbf{u}_+ - \mathbf{u}_-)/\sqrt{\mu}, \tag{18}$$

and the b's are force constants. Considering periodic solutions and eliminating **w** in Eqs. (17), the dielectric function is obtained

$$\varepsilon(\omega) = 1 + 4\pi b_{22} - \frac{4\pi b_{12} b_{21}}{b_{11} + \omega^2}. \tag{19}$$

The b_{ij}'s can be expressed in terms of measurable quantities

$$b_{11} = -\omega_{TO}^2, \qquad b_{12} = b_{21} = \left(\frac{\varepsilon_0 - \varepsilon_\infty}{4\pi}\right)^{1/2} \omega_{TO}, \qquad b_{22} = \frac{\varepsilon_\infty - 1}{4\pi}, \tag{20}$$

where ω_{TO} is the frequency of the transverse optical mode at $\mathbf{q} = 0$, ε_0 is the static dielectric constant, and ε_∞ the high-frequency dielectric constant.

The solution of Eqs. (17) simultaneously with Maxwell's equations represents the coupled phonon–photon system known as polaritons. For an infinite crystal this yields the usual optical modes. In the case of finite crystals, certain of the solutions are localized in one or more directions. For example, in a semi-infinite crystal, if there is no free charge

$$\nabla \cdot \mathbf{D} = 0, \tag{21}$$

and the normal component of **D** is continuous across the boundary. Also

$$\mathbf{E} = -\nabla V, \tag{22}$$

with V continuous across the boundary. Solutions for this case are of the form

$$V = V_0 \exp(-q_z z) \cos(\mathbf{q} \cdot \mathbf{r}_{\parallel} - \omega t). \tag{23}$$

The corresponding electric field represents a surface wave that has associated with it a field in the vacuum. This is shown schematically in Fig. 6. The

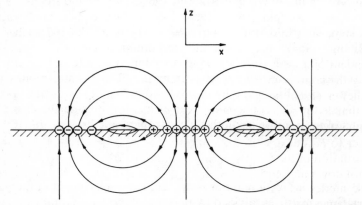

Fig. 6. Schematic of the electric field of a surface wave in a semi-infinite optically active isotropic material. The field penetrates into the vacuum. The charged spheres indicate instantaneous regions of net positive and negative charge as the ionic sublattices vibrate against each other.

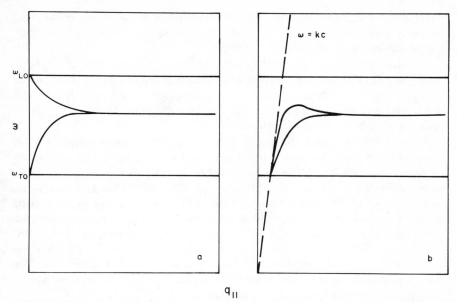

Fig. 7. Dispersion relation for the Fuchs–Kliewer phonon (a) without retardation and (b) with retardation.

frequency of the wave is determined from Eqs. (19), (20), and (21) and is

$$\omega^2 = \frac{\varepsilon_0 + 1}{\varepsilon_\infty + 1}\omega_{\text{TO}}^2.$$ (24)

This is somewhat lower than the bulk longitudinal optical phonon (Born and Huang, 1954) and in the electrostatic approximation $(c \to \infty)$ is independent of $\mathbf{q}$.

Using the continuum model and the electrostatic approximation, Fuchs and Kliewer (1965) were the first to calculate modes of this kind. They found that as $\mathbf{q}_{\parallel} \to 0$ the penetration of the modes gets large, similar to Rayleigh waves, and that the modes split in frequency, one going to ω_{LO} and the other to ω_{TO} at $\mathbf{q} = 0$. These results are strongly affected by including the retardation of the Coulomb interaction $[(1/c)(\partial H/\partial t) \neq 0]$. Both modes then turn down and go to ω_{TO} at $\mathbf{q}_{\parallel} \neq 0$. This is illustrated in Fig. 7.

These modes, as well as others with shallow penetration at $\mathbf{q}_{\parallel} = 0$ due to short-range forces near the surface (Lucas, 1968), have been found also in lattice-dynamics calculations (Tong and Maradudin, 1969; Chen et al., 1971b), using the Kellerman rigid-ion model (Kellerman, 1940), which neglects retardation and thus is the analog of the electrostatic approximation in the continuum model. Additionally, of course, a series of optical surface modes at short wavelengths are found, which are similar to the modes found for monatomic crystals.

In calculations using the shell model, which, unlike the rigid-ion model, takes the polarizabilities of the ions into account, Chen *et al.*, (1971b, 1972b, c) find that the Fuchs–Kliewer modes appear as part of the bulk bands and are nowhere pure surface modes, and that they therefore do not carry much information about the surface per se.

Nevertheless, the fact that there is a dipole field associated with these modes provides the possibility of measuring their frequency and dispersion by interaction either with an electromagnetic field or a charged particle. Because the wave vectors of photons at the relevant frequencies are of the order of 10^{+3} cm^{-1}, only phonons with small wave vectors can be excited by direct processes. The optical absorption by surface phonons and its crystal size and shape dependence have been discussed (Ruppin and Engleman, 1970; Ibach 1971a). Flat crystals have only nonradiative surface modes ($\omega > kc$) and therefore show no direct absorption due to surface phonons. Nonradiative surface phonons have been observed, however, by Attenuated Total Reflection (Marschall and Fischer, 1972; Bryksin *et al.*, 1972; Barker, 1972), essentially by using a prism to allow coupling of the beam to the surface phonon. An example of the dispersion relation at small $q_{||}$ obtained for GaP is shown in Fig. 8.

Optically active phonons have been seen by scattering of fast electrons (Boersch *et al.*, 1968; Chase and Kliewer, 1970) and slow electrons (Ibach, 1970, 1971b). Because of the sensitivity of slow electrons to surfaces and their possible use for determining phonon spectra for large $q_{||}$, the low-energy electron spectroscopy measurements will be discussed more fully in Section VII.

Several other experimental methods have been used to detect optically active surface phonons. Evans *et al.* (1973) recognized that a material of high dielectric constant in contact with a crystal would depress the surface phonon dispersion curve for the crystal sufficiently below the bulk LO phonon frequency to make it observable with Raman scattering. They grew a thin film of (111) GaAs on a (0001) sapphire substrate in order to obtain the surface polariton dispersion curve for GaAs. Lüth (1972) has detected the Fuchs–Kliewer phonons in oscillatory photoconductivity measurements on ZnO.

In addition to the measurements of optically active modes described above, some indirect evidence of the effect of surface phonons is available from several others experiments. Rieder and Hörl (1968) have measured the

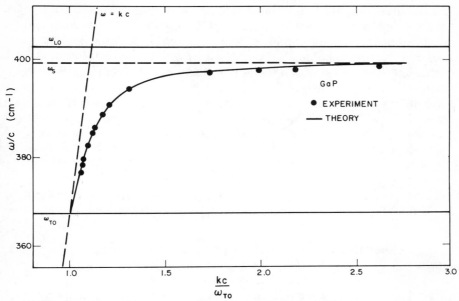

Fig. 8. Dispersion relation for surface polaritons in GaP using ATR. ω_{LO} and ω_{TO} are the longitudinal and transverse bulk optical mode frequencies respectively, and ω_s is the limiting frequency of the surface mode. (After Marschall and Fischer, 1972.)

frequency distribution of inelastically scattered neutrons from small crystallites of MgO. They ascribe peaks in the difference of the frequency distribution for small grains and a macroscopic crystal to the excitation of surface phonons. Such interpretations must be viewed with caution, however, because of the possibility of changes in the bulk phonon spectrum as the crystallite size is made very small, as pointed out earlier.

In experiments to measure the change in the superconducting transition temperature with the adsorption of noble-gas overlayers on thin-film substrates, Naugle *et al.* (1973) have found evidence of a surface phonon contribution to thin-film superconductivity. The contribution of surface phonons to the determination of T_c is small, and changes in T_c can be caused by a number of other factors. The agreement between their experimental results and a simple calculation of the modification of the phonon spectrum by a physisorbed layer, however, lead Naugle *et al.* to suggest that the dominant factor in the change in T_c is the surface phonons.

Finally, surface phonons make a considerable contribution to the thermal diffuse scattering (TDS) of low-energy electrons and may be important in the inelastic scattering of atoms from surfaces. In the former, it has been estimated (Huber, 1967) that surface phonon scattering may account for as much as half of the TDS intensity already at very small $\mathbf{q}_\parallel$ values. Because of the limited energy resolution of low-energy electron diffraction systems and the fact that phonons with a range of $q_\perp$ contribute to the diffuse scattering at every value of $\mathbf{q}_\parallel$, it is not possible to determine the phonon dispersion relations directly. However, because of the large surface phonon contribution, comparisons with calculated TDS profiles for assumed phonon distributions should make this a useful tool for determining surface phonon spectra. Thermal diffuse scattering will be discussed in more detail in Section VII.

IV. VIBRATIONAL CONTRIBUTIONS TO SURFACE THERMODYNAMIC FUNCTIONS

In Section II it was remarked that the creation of a surface will add to the total energy of the crystal a term proportional to the surface area, even in the absence of temperature effects, because of the static relaxation of the crystal to maintain mechanical equilibrium. Inclusion of vibrational contributions changes the energy in two ways. Due strictly to the thermal expansion of the lattice, there will be a temperature dependence to Φ^s in Eq. (1). Both $\Phi_0^s(T)$ and $\Delta\Phi^s(T)$ may change, the first because of bulk thermal expansion and the latter because of excess surface thermal expansion. In addition there will be a term depending on the vibrational frequency spectrum. Thus the total surface energy E^s at a temperature T is

$$E^s(T) = \Phi^s(T) + E^s_{\text{vib}}(T), \tag{25}$$

where $\Phi^s(T)$ includes all the static effects and $E^s_{\text{vib}}(T)$ represents the energy of the vibrational motion of surface atoms about their lattice sites, in excess of the bulk. This term therefore depends on the normal mode spectrum.

Similarly there will be surface vibrational contributions to other thermodynamic functions. In the harmonic approximation, the lattice contribution to a thermodynamic property P_{vib} is the expectation value of a function $p(x)$,

$$P_{\text{vib}} = \int_0^{\omega_{\text{max}}} p(x) f(\omega) \, d\omega, \tag{26}$$

where $f(\omega)$ is the frequency distribution function,

$$x = \frac{\hbar\omega_n(\mathbf{q}_\parallel)}{kT}, \tag{27}$$

and ω_n is the frequency of the nth mode with wave vector $\mathbf{q}_{\parallel}$. The form of $p(x)$ for the energy, free energy, entropy, and specific heat is, respectively,

$$p_E(x) = kT\{\tfrac{1}{2}x + [x(\exp(x)-1)]\}, \tag{28}$$

$$p_F(x) = kT\{\tfrac{1}{2}x + \ln[1-\exp(-x)]\}, \tag{29}$$

$$p_S(x) = k\{-\ln[1-\exp(-x)] + [x/(\exp(x)-1)]\}, \tag{30}$$

$$p_{C_V}(x) = k\{x^2\exp(x)/[\exp(x)-1]^2\}. \tag{31}$$

The determination of surface thermodynamic functions thus reduces to calculating the surface frequency distribution $f^s(\omega)$ and integrating Eq. (26). Since the difference between $f(\omega)$ for a crystal with and without surfaces is small, and since thermodynamic functions involve averages of frequency distributions, the surface contributions are small unless the surface area–to–volume ratio becomes very large (e.g., thin films), and are not very sensitive to details of the phonon spectrum.

A particularly simple order-of-magnitude estimate of the vibrational parts of surface thermodynamic functions and their temperature dependence is obtained with a continuum model assuming a Debye spectrum of modes that includes a term proportional to the surface area

$$f(\omega) = \alpha V\omega^2 + \beta A\omega, \tag{32}$$

where α and β are constants depending on the elastic characteristics of the medium. Continuing to limit the total number of modes to $3N$ then just changes ω_{max} in Eq. (26). This method was applied by Benson and Yun (1967) to rare-gas and ionic crystals at $0°$ K and the melting temperature. They obtain, for example, a vibrational contribution of $-2.4\,\text{ergs/cm}^2$ to the surface energy of Ne at $0°$ K, which is to be compared to a static surface energy of 19.7 ergs/cm^2. At the melting point the vibrational contribution is much less negative. Because of the Debye approximation the values are independent of surface orientation.

It is easy to see why the vibrational surface energy should be negative. The frequency distribution of the crystal with surfaces is weighted toward low frequencies in relation to the infinite crystal because the surface causes a softening of the lattice vibrations, as shown in Section III. The shift is proportional to surface area. Hence for all temperatures, $E_{vib}^s(T)$ will be negative. That it approaches zero for high temperatures is reasonable because in this limit each vibrational mode contributes kT of energy to the total energy.

Lattice-dynamical calculations, although less transparent, allow a more detailed description of the vibrational contribution to thermodynamic functions. In particular, thermal expansion and changes in force constants

can be more easily taken into account. Using the quasiharmonic approximation and the Lennard–Jones potential, Allen and de Wette (1969b) calculate
the vibrational parts of the surface thermodynamic functions by evaluating
Eq. (26) using the normal mode distribution of a crystal with and without
surfaces. Then

$$P_{\text{vib}}^{\text{s}}(T) = (N/A)[p(T) - p^{\text{b}}(T)], \tag{33}$$

where $p^{\text{b}}(T)$ is the value of the thermodynamic function per particle for the
crystal with cyclic boundary conditions and $p(T)$ includes the surfaces. N
is the number of particles in the crystal and A is the surface area. Some
results for the (111) surface of a noble-gas solid including bulk thermal
expansion but no excess surface thermal expansion are shown in Figs. 9–11
as a function of the temperature. The values for $E_{\text{vibr}}^{\text{s}}$ (Fig. 9), as already
mentioned, approach zero from negative values. The values calculated by
Benson and Yun (1967) at 0° K and the melting point are also shown.
Figure 10 shows the surface entropy $S^{\text{s}}(T)$. At 0° K, S^{s} must be zero; at
high temperatures it is nearly constant, because here it is proportional to
the difference of the logarithm of the frequency distributions for the
crystal with and without surfaces. These are the same except for the small
effect of thermal expansion. If excess surface thermal expansion were taken
into account, S^{s} would decrease slightly from its maximum value at very
high temperatures. Between the high- and low-temperature limits S^{s} changes

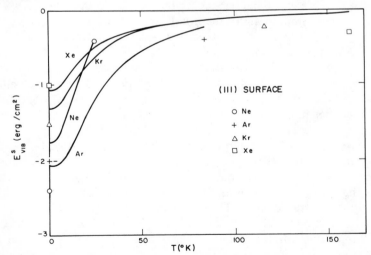

Fig. 9. Temperature dependence of the vibrational contribution to the surface energy for
Lennard–Jones solids with a (111) surface. The symbols at 0° K and the melting points are
values calculated by Benson and Yun (1967) for a Debye model. (After Allen and de Wette,
1969b.)

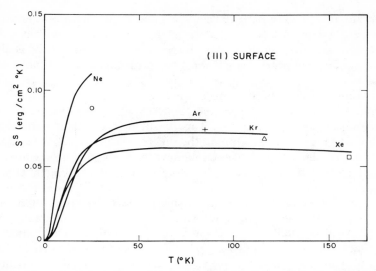

Fig. 10. Temperature dependence of the surface entropy for Lennard–Jones solids with a (111) surface. The four symbols are values from Benson and Yun for a Debye model. (After Allen and de Wette, 1969b.)

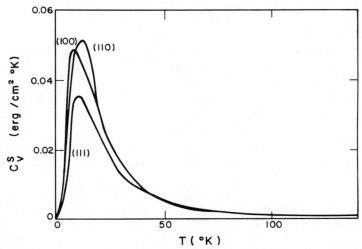

Fig. 11. Temperature dependence of the surface contribution to the specific heat as a function of surface orientation for Xe. (After Allen and de Wette, 1969b.)

rapidly. Figure 11 shows the surface contribution to the specific heat and its dependence on surface orientation. c_v^s must be zero at $0°$ K and must approach zero at high temperatures for the same reason E_{vibr}^s does. The peak at intermediate temperatures indicates the presence of the surface modes; c_v^s rises approximately like T^2 at low temperatures,[†] and, reflecting the change in vibrational surface energy shown in Fig. 9, reaches a maximum and falls to zero. Although qualitatively the same, c_v^s values for different surface orientations differ quantitatively. This anisotropy is due to the different coordination number of an atom in the different surfaces; the (111) surface, being the most densely packed and retaining the most neighbors, contributes less surface vibrational energy to the total energy of the crystal than the other faces.

Methods to measure surface thermodynamic functions have been reviewed elsewhere (Benson and Yun, 1967; Maradudin *et al.*, 1971). The methods have been mostly macroscopic, under conditions that were in part un-controllable, and it has not been possible to ascribe measured values in excess of bulk thermodynamic quantities conclusively to the contribution of the clean surface. So, for example, cleavage experiments must account for the energy of plastic deformation and, unless carried out in ultrahigh vacuum, are further unreliable as to the mechanism responsible for the excess quantities. Roughness may affect the surface area determination, as well as introduce unknown surface orientations, in calorimetric experiments on fine powders. Very small particles may themselves have vibrational and rotational energy that can contribute to the specific heat. Nevertheless, qualitative agreement between experiment and theory has been obtained in at least one case, where the temperature of the peak in c_v^s for NaCl (Morrison and Patterson, 1956) agrees with theory (Chen *et al.*, 1971b).

Finally, there is the effect of finite size of particles. As mentioned earlier, the bulk modes at long wavelengths and low frequencies are changed for very small particles or very thin films. Since these modes contribute most to the low-temperature specific heat, a decrease in the size of the particle should lead to a decrease in the specific heat. In a recent careful study of the specific heat of very small lead and indium particles impregnated into porous glass, Novotny and Meincke (1973) found evidence for this effect. An excess heat capacity is found, which at the higher temperatures ($15°$ K) is in fair agreement with predictions of the Debye model, Eq. (32), for the surface specific heat. However, comparison of the temperature dependence of this surface specific heat with the Debye model indicates a weaker than T^2 dependence at low temperatures. Since models for thick crystals that

---

[†] For materials for which dispersive and anisotropy effects are already important at long wavelengths and low frequencies, the temperature dependence differs from the Debye-like T^2.

take relaxation and anharmonic effects into account predict a stronger than T^2 dependence, this result is ascribed to the effect of finite size on the bulk phonon spectrum.

Whereas direct measurements of surface thermodynamic functions seem to be quite difficult, it would be possible to obtain an estimate of their magnitude if the surface frequency distribution function were known from other experiments. We turn now to a discussion of the mean square vibrational amplitudes of surface atoms, which also depend on the average over the frequency function but which are much more easily measurable.

V. MEAN SQUARE AMPLITUDE OF VIBRATION

The displacement of any atom in the lattice can be expressed as the sum of its displacements due to a number of waves having various amplitudes and wavelengths. The presence of a surface introduces some new waves. Hence it is reasonable that the mean square displacement of atoms near the surface should be different from that of bulk atoms. The simple model of a harmonic oscillator with force constant σ serves to illustrate the general properties of the rms vibrational amplitude of surface atoms. For pairwise interactions

$$\langle u^2 \rangle = kT/\sigma; \qquad (34)$$

thus the smaller the force constant, the larger the mean square displacement. If, for a surface atom, half the neighbors are missing, $\sigma = \frac{1}{2}\sigma_{\text{bulk}}$, and $\langle u^2 \rangle$ normal to the surface will be twice the bulk value. Similarly, for directions parallel to the surface, fewer neighbors contributing to the force constant are missing, and one expects a smaller excess vibrational amplitude. Finally, as one goes to deeper layers, σ begins to approach the bulk value, so the mean square amplitudes will approach the bulk values. This decay occurs in a depth comparable to the shortest wavelength phonon in the crystal.

The mean square displacement of surface atoms is a quantity readily accessible in scattering experiments and thus has attracted a good deal of attention. As mentioned above, even in the absence of anharmonic effects, simply the reduced symmetry of an atom at a surface will lead to anisotropy in $\langle u^2 \rangle$ and magnitudes differing from the bulk. In addition, of course, anharmonic effects at higher temperatures as well as force–constant changes (e.g., by adsorption of a gas) will affect the mean square amplitudes. We will discuss the experimental determination of $\langle u^2 \rangle$ in detail in Section VI; here we concentrate on the model calculations.

Most determinations of $\langle u^2 \rangle$ have been made for a discrete lattice using the harmonic or quasiharmonic approximation. For this model the displacement in the α direction for the Kth atom in the lth unit cell is given by Eqs. (11) and (12). From this one can obtain (Maradudin et al., 1971) a solution for the mean square amplitude $\langle u_\alpha^2(l_3, K) \rangle$, which is the thermal average of u_α for the Kth atom in the l_3th plane,

$$\langle u_\alpha^2(l_3, K) \rangle = \frac{\hbar}{2L^2 M_K} \sum_{\mathbf{q}_\parallel, n} |\xi_\alpha(l_3, K; \mathbf{q}_\parallel, n)|^2 \coth\left(\frac{\hbar \omega_n(\mathbf{q}_\parallel)}{2kT}\right) \Big/ \omega_n(\mathbf{q}_\parallel), \qquad (35)$$

or in terms of the dynamical matrix of Eq. (14),

$$\langle u_\alpha^2(l_3, K) \rangle = \frac{\hbar}{2M_K} \left[\tilde{D}^{-1/2} \coth\left(\frac{\hbar \tilde{D}^{1/2}}{2kT}\right) \right]_{l_3, \alpha; l_3, \alpha}. \qquad (36)$$

L^2 is the number of atoms in a plane. $|\xi_\alpha|^2$ gives the amplitude of vibration at the l_3th plane due to the wave with frequency $\omega_n(\mathbf{q}_\parallel)$. Thus the rms vibrational amplitude of each atom layer may be determined separately. In the low-temperature limit, Eq. (36) reduces to

$$\langle u_\alpha^2(l_3, K) \rangle = \frac{\hbar}{2M_K} [\tilde{D}^{-1/2}]_{l_3, \alpha; l_3, \alpha}, \qquad (37)$$

which specifies the zero-point motion, and in the high-temperature limit,

$$\langle u_\alpha^2(l_3, K) \rangle = \frac{kT}{M_K} [\tilde{D}^{-1}]_{l_3, \alpha; l_3, \alpha}, \qquad (38)$$

which corresponds to Eq. (34) for coupled oscillators.

Eq. (36) has been evaluated for a range of model crystals and surface orientations and for various approximations. Clark et al. (1965) calculated the mean square vibrational amplitudes for several surfaces of a nearest-neighbor, central-force model of an fcc lattice in the high-temperature limit. Surface force constants were taken identical to the bulk. As expected from the above discussion of the harmonic oscillator model, the rms vibrational amplitudes normal to the surface were about twice those in the bulk, whereas the parallel components were nearer bulk values and isotropic or not, depending on the surface orientation. A decay of $\langle u_\perp^2 \rangle$ to bulk values within five planes was found. Because of its assumption of central forces, this model is too simple to be compared to measurements for most materials. However, the bulk properties of Ni appear to be fairly well described by a central-forces model. Comparison of these calculations with measurements of the *effective* mean square vibrational amplitude of atoms in a Ni(110) surface (MacRae, 1964) indicates that the theoretical values are too low. This is not surprising in view of the neglect of surface force–

constant changes and anharmonic effects, which act to make the surface vibrational amplitude larger. In fact, the discrepancy is considerably greater than indicated, because the actual surface rms amplitude must be larger than the effective $\langle u_\perp^2 \rangle$ measured by low-energy electron diffraction experiments (see Section VI). This seems to have been almost universally neglected in comparisons of theory and experiment.

Wallis *et al.* (1968) modified the above model by changing the force constants near the surface. However, to improve the agreement between theory and experiment required, in addition to a reasonable weakening of force constants between planes, an unreasonably large increase in the coupling to other atoms in the surface plane. Vail (1967), using a Morse potential, indicates a slight stiffening of these force constants may be possible; here a factor of 2.2 is required.

Calculations with the nearest-neighbor-forces model on diamond-structure crystals have been made using a different approximation (Theeten and Dobrzynski, 1972). It was assumed that $\langle u^2 \rangle$ could be satisfactorily described by using only the first two terms of a Taylor expansion of the RHS of Eq. (36). The first term is just the Einstein term, Eq. (34), for uncoupled oscillators. Hence this approximation is equivalent to weak coupling and is better at low temperatures than high. Figure 12 demonstrates the differences in $\langle u^2 \rangle$ as a function of the temperature resulting from use of an Einstein oscillator model and inclusion of the next two terms in the Taylor expansion.

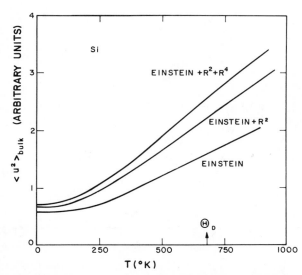

Fig. 12. Temperature dependence of the bulk mean square vibrational amplitude in Si for Einstein oscillators and first two coupling terms. (After Theeten and Dobrzynski, 1972.)

Force–constant models exhibit the qualitative features of the vibrational behavior of atoms at the surface but cannot be expected to be quantitative without inclusion of anharmonic effects. As already mentioned, such effects are much more easily included in calculations using model potentials. Unfortunately, most measurements so far have been performed on metals, which are hard to model. Most calculations have been performed with Lennard–Jones or Morse potentials. Morse potential calculations have been performed in the Einstein approximation, Eq. (34) (Jackson, 1974). The calculations using the Lennard–Jones potential (Allen and de Wette, 1969a, c) are more realistic. Since this potential describes the noble-gas solids quite well, and since a recent measurement lends some basis for comparison, we will describe them briefly.

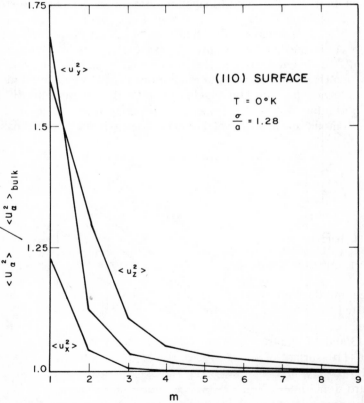

Fig. 13. Ratios $\langle u_\alpha^2 \rangle / \langle u_\alpha^2 \rangle_{\text{bulk}}$ at $T = 0°$ K as a function of depth into the crystal for the (110) surface of a Lennard–Jones solid with density corresponding approximately to Ar. (After Allen and de Wette, 1969a.)

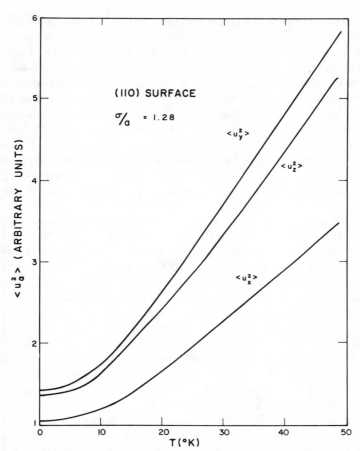

Fig. 14. Temperature dependence of $\langle u_{\alpha}^2 \rangle_{\text{surf}}$ for the (110) surface of a Lennard–Jones solid with density corresponding approximately to Ar. For this density, the Debye temperature $\theta_{\text{D}} \simeq 92°$ K. (After Allen and de Wette, 1969a.)

Allen and de Wette studied the effect of various approximations on the calculated $\langle u^2 \rangle$ values. Some results using the quasiharmonic approximation, i.e., including static but not dynamic displacements, are shown in Figs. 13 and 14. Figure 13 shows the mean square amplitude of vibration for the (110) surface of an fcc solid at $0°$ K and its decay with depth. Even at $0°$ K, $\langle u^2 \rangle_{\text{surf}}$ in some directions is almost twice the bulk value. If the Lennard–Jones interaction is cut off at nearest neighbors only, these values drop by about 25%. Figure 14 shows the temperature dependence of these amplitudes. As expected for the harmonic approximation, $\langle u_{\alpha}^2 \rangle$ is linear with temperature in the high-temperature limit. Linearity down to about one half the Debye temperature was found earlier by Wallis *et al.* (1968). If dynamic displacements (i.e., bulk thermal expansion) are included, $\langle u_\alpha^2 \rangle$ rises faster than linearly. Of more interest is the ratio $\langle u_\alpha^2 \rangle_{\text{surf}} / \langle u_\alpha^2 \rangle_{\text{bulk}}$. This quantity increases sharply in the range between the low- and high-temperature limits. This can be understood by referring again to the normal mode spectrum. At $0°\text{K}\langle u_\alpha^2 \rangle$ is proportional to $1/\langle \omega \rangle$ and at high temperatures it is proportional to $1/\langle \omega^2 \rangle$, indicated by Eqs. (37) and (38). Hence as the temperature is raised, the lower frequencies of the surface modes are weighted more heavily and $\langle u_\alpha^2 \rangle_{\text{surf}}$ gains over $\langle u_\alpha^2 \rangle_{\text{bulk}}$. By a similar argument it is clear that bulk thermal expansion (which changes only the density of the crystal), although it affects the values of $\langle u_\alpha^2 \rangle$, has little effect on the ratio $\langle u_\alpha^2 \rangle_{\text{surf}} / \langle u_\alpha^2 \rangle_{\text{bulk}}$, since a change in density shifts all the modes.

Truly anharmonic surface effects become important in Lennard–Jones solids at about half the melting temperature. The change in the effective mean position of surface atoms (static displacement) caused by the asymmetry of forces at the surface should be strongly temperature-dependent and should give an excess surface thermal expansion. Such anharmonic effects were included in a determination of $\langle u^2 \rangle$ by Allen *et al.* (1969b) using the method of molecular dynamics. This method consists of integrating Newton's equations of motion for a system of interacting particles until they reach thermal equilibrium. Thus anharmonic effects are taken into account completely for a given interaction potential. The method is valid for temperatures above Θ_D, where zero-point effects no longer play a role. A comparison of results obtained with molecular dynamics and lattice dynamics at half the melting temperature is given in Fig. 15. The lattice dynamics calculation was made in the high-temperature limit using dynamic displacements. Deep in the crystal the results are similar, but near the surface the effect of anharmonicity is evident. Mean square vibrational amplitudes for all directions are larger. Table I summarizes the results of the various models. Two lattice-dynamics calculations in the high-temperature limit and the molecular-dynamics method are compared. As the model becomes more realistic, $\langle u_\alpha^2 \rangle_{\text{surf}} / \langle u_\alpha^2 \rangle_{\text{bulk}}$ ratios become increasingly larger. The large influence of anharmonic effects can be understood by considering the asymmetric form of the Lennard–Jones potential. If one considers the motion of the surface plane compared to one in the bulk, one sees that the "effective spring constant" for the surface plane gets smaller as the distance from the lower plane increases. This results in static displacements as well as a strong dependence of the equilibrium lattice constant on temperature.

So far only one measurement has been made on noble-gas solids (Ignatiev and Rhodin, 1973). From Debye–Waller factor measurements of low-energy electron diffraction intensities, an effective surface-to-bulk ratio

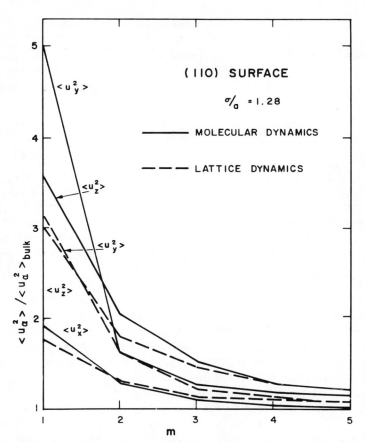

Fig. 15. Comparison of layer dependence of the mean square vibrational amplitude at half the melting temperature calculated including all anharmonic effects (molecular dynamics) and in the quasiharmonic approximation. (After Allen *et al.*, 1969b.)

of about 2 of mean square vibrational amplitudes perpendicular to the (111) surface of Xe is found. Comparison of the data with models giving $\langle u_\perp^2 \rangle_{\text{surf}} / \langle u_\perp^2 \rangle_{\text{bulk}} = 2.25$ (lattice dynamics) and 3.48 (molecular dynamics) and including the effect of penetration of the electron beam into the crystal indicates that both models fit the data adequately (Tong *et al.*, 1973). The measurements are not accurate enough to differentiate between the models. These results will be discussed in greater detail in Section VI. However, a considerable number of measurements of $\langle u^2 \rangle_{\text{surf}}$ on metals exist. These are frequently on the high side of even the last column in Table I, even though they are usually taken at temperatures below half the melting temperature. It is, of course, not expected that the Lennard–Jones potential will be

TABLE I

The Ratios $\langle u_\alpha^2 \rangle_{\text{surf}} / \langle u_\alpha^2 \rangle_{\text{bulk}}$ for an FCC Lennard–Jones Solid Calculated in the High-Temperature Limit Approximation[a]

Surface	Component	SFC[b]	QH, $T_m/2^c$	MD, $T_m/2^d$
(100)	$\langle u_x^2 \rangle = \langle u_y \rangle^2$	1.46	2.03	2.23 ± 0.17
	$\langle u_z^2 \rangle$	1.87	2.77	3.07 ± 0.15
(111)	$\langle u_x^2 \rangle = \langle u_y^2 \rangle$	1.30	1.45	1.27 ± 0.07
	$\langle u_z^2 \rangle$	1.86	2.85	3.48 ± 0.18
(110)	$\langle u_x^2 \rangle$	1.50	1.78	1.89 ± 0.20
	$\langle u_y^2 \rangle$	2.14	3.21	4.99 ± 0.65
	$\langle u_z^2 \rangle$	1.83	3.03	3.57 ± 0.25

[a] After Allen and de Wette (1969c).
[b] Simple force-constant model, with force constants at the surface equal to those in bulk.
[c] Quasi-harmonic approximation, changes in surface force constants determined at $\frac{1}{2}$ the melting temperature.
[d] Molecular dynamics computer experiment at $T_m/2$, with all effects including anharmonicity taken into account.

quantitatively accurate for metals. However, a second point may be of equal importance. So far the possibility that the pairwise interatomic potential may depend on the orientation and depth of the pair relative to the surface has not been considered in the calculations. The likelihood of this is indicated by recent experiments by Unertl (1973). It could, of course, drastically alter the surface–to–bulk vibrational amplitude ratio.

As already noted, the mean square amplitudes represent averages over all vibrational modes and should thus not be particularly sensitive to the details of the vibrational spectrum. Nevertheless, it would be interesting to see what the relative contributions of the four different types of modes listed in Section III.A are. Some simple statements can be made without analysis. From Eq. (35) it is clear that modes with lower frequency contribute more to $\langle u_\alpha^2 \rangle$. Thus in diatomic materials the acoustic surface modes are more important than optical surface modes. Further, if the acoustic modes themselves are relatively high in frequency, the corresponding $\langle u_\alpha^2 \rangle$ will be low. This is borne out in calculations for NaCl (Chen et al., 1972a), for which the acoustic surface modes are very close to the bulk and the ratio $\langle u_\perp^2 \rangle_{\text{surf}} / \langle u_\perp^2 \rangle_{\text{bulk}}$ is only 1.6.

To determine what the relative contribution is to $\langle u_\alpha^2 \rangle$ of the several types of modes in a monatomic solid, Dennis and Huber (1972) used the continuum model to evaluate the surface mean square amplitudes. Despite its limitations, this model allows a graphic representation of the

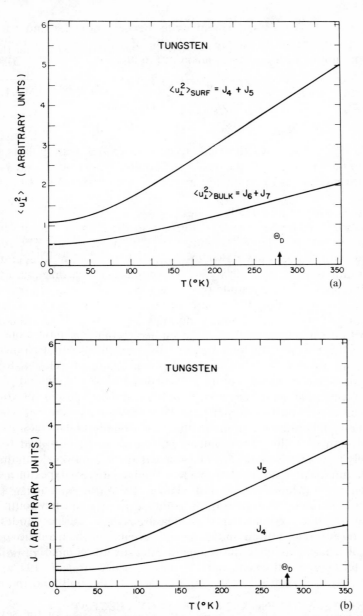

Fig. 16. Contribution of various types of modes to the mean square vibrational amplitude normal to the surface. (a) Temperature dependence of $\langle u_\perp^2 \rangle_{surf} = J_4 + J_5$ and $\langle u_\perp^2 \rangle_{bulk} = J_6 + J_7$ for an isotropic continuum representative of W. (b) Contribution of pure surface modes (J_5) and mixed surface–bulk modes (J_4) to $\langle u_\perp^2 \rangle_{surf}$. (After Dennis and Huber, 1972.)

makeup of $\langle u_\alpha^2 \rangle$. Dennis and Huber write the perpendicular and parallel components of the mean square amplitude of vibration of an isotropic elastic continuum with a stress-free surface as the sum of several types of modes.

$$\langle u_\parallel^2 \rangle_{\text{surf}} = C[J_1(T) + J_2(T) + J_3(T)] \qquad (39\text{a})$$

$$\langle u_\perp^2 \rangle_{\text{surf}} = C[J_4(T) + J_5(T)] \qquad (39\text{b})$$

$$\langle u_\parallel^2 \rangle_{\text{bulk}} = \langle u_\perp^2 \rangle_{\text{bulk}} = C[J_6(T) + J_7(T)]. \qquad (39\text{c})$$

In Eq. (39a), $J_1(T)$ are the bulk transverse modes with polarization parallel to the surface, $J_2(T)$ the mixed and combination modes, and $J_3(T)$, the surface modes. $J_4(T)$ and $J_5(T)$ in Eq. (39b) correspond to $J_2(T)$ and $J_3(T)$, respectively, in Eq. (39a). $J_6(T)$ and $J_7(T)$ are, respectively, the transverse and longitudinal bulk modes.

Results for $\langle u_\perp^2 \rangle$ for W, which is isotropic in the long-wavelength limit, are shown in Fig. 16. Figure 16a shows $\langle u_\perp^2 \rangle_{\text{surf}}$ and $\langle u_\perp^2 \rangle_{\text{bulk}}$. The curves are linear with temperature down to about half the Debye temperature, and their ratio is about 2.5. Below $\frac{1}{2}\Theta_D$, zero-point effects become important. Figure 16b shows the magnitude of the two types of modes contributing to $\langle u_\perp^2 \rangle_{\text{surf}}$. It is clear that the pure surface modes are responsible for more than half of $\langle u_\perp^2 \rangle_{\text{surf}}$. A similar analysis indicates that the transverse modes with polarization parallel to the surface account for more than half of $\langle u_\parallel^2 \rangle_{\text{surf}}$ in the limit of long wavelengths. Anything that significantly alters the surface modes is thus likely to have a strong effect on the mean square vibrational amplitudes of the surface plane.

Several techniques have been used to determine mean square vibrational amplitudes. A few measurements have been made using the Mössbauer effect. The results are ambiguous in that occasionally surface vibrational amplitudes smaller than bulk values are measured. These studies are adequately reviewed elsewhere (Goldanskii and Suzdalev, 1971; Maradudin, 1966). Measurements using grazing incidence high-energy electron diffraction have been made (Menzel–Kopp and Menzel, 1956). Recently it has been shown that atomic scattering from solid surfaces may be useful for obtaining $\langle u^2 \rangle$ (see Section VIII). Most measurements of surface vibrational effects, however, have been made using low-energy electron scattering because of the sensitivity of the scattered intensity to thermal vibrations. We turn to these measurements next.

VI. ELASTIC SCATTERING OF LOW-ENERGY ELECTRONS

Most all reliable experimental information on surface atom vibrations has come from the scattering of low-energy electrons. It is natural that slow electrons should play such a dominant role. Of all possible probes, they are the simplest that still interact sufficiently strongly with matter to be sensitive to the outermost few atomic planes of a crystal. Because of the large thermal amplitudes of these atoms, temperature effects are usually quite prominent in low-energy electron scattering. Both elastic diffraction (LEED) and inelastic scattering have produced new insights into the dynamical behavior of surface atoms. In this section the influence of surface atom vibrations on elastic scattering will be discussed; Section VII deals with the phonon-assisted, inelastic scattering.

A. Low-Energy Electron Diffraction

Experimental as well as theoretical aspects of low-energy electron diffraction have been reviewed frequently [for a list of references, see Chapter 1 of Volume I and also Webb and Lagally (1973)]. We briefly indicate here the necessary background for surface lattice–dynamics measurements. For penetrating radiation incident on a lattice of identical scatterers, the elastically scattered intensity can be written in the kinematic (single-scattering) approximation (James, 1965) as

$$I(\mathbf{S}) = |f(2\theta, E)|^2 \sum_{i,j} \exp[i\mathbf{S} \cdot (\mathbf{r}_i + \mathbf{u}_i - \mathbf{r}_j - \mathbf{u}_j)], \tag{40}$$

where $\mathbf{S} = \mathbf{k} - \mathbf{k}_0$ is the momentum transfer or diffraction vector, $f(2\theta, E)$ is the atomic scattering factor, $\mathbf{r}_i$ is the equilibrium position of the ith atom, and $\mathbf{u}_i$ is its displacement. Because in typical low-energy electron diffraction systems (as for x rays) the energy resolution is insufficient to resolve phonon energies, the diffraction experiment is equivalent to scattering from instantaneous and stationary configurations of scatterers.

Because the motion of atoms is rapid compared to the time of a measurement, a thermal average of Eq. (40) is required,

$$\langle I(\mathbf{S}) \rangle = |f(2\theta, E)|^2 \sum_{ij} \exp[i\mathbf{S} \cdot (\mathbf{r}_i - \mathbf{r}_j)] \langle \exp[i\mathbf{S} \cdot (\mathbf{u}_i - \mathbf{u}_j)] \rangle. \tag{41}$$

Assuming the harmonic approximation is valid, this can be written as

$$\langle I(\mathbf{S}) \rangle = |f(2\theta, E)|^2 \exp[-\langle (\mathbf{S} \cdot \mathbf{u})^2 \rangle] \sum_{ij} \exp[i\mathbf{S} \cdot (\mathbf{r}_i - \mathbf{r}_j)]$$

$$\times \{1 + \langle (\mathbf{S} \cdot \mathbf{u}_i)(\mathbf{S} \cdot \mathbf{u}_j) \rangle + [\exp[\langle (\mathbf{S} \cdot \mathbf{u}_i)(\mathbf{S} \cdot \mathbf{u}_j) \rangle] - 1 - \langle (\mathbf{S} \cdot \mathbf{u}_i)(\mathbf{S} \cdot \mathbf{u}_j) \rangle]\}. \tag{42}$$

The factor $\exp[-\langle \mathbf{S} \cdot \mathbf{u})^2 \rangle] = \exp[-2M]$ is the Debye–Waller factor, which determines the temperature dependence of the Bragg or zero-phonon scattering,

$$\langle I^{(0)}(\mathbf{S}) \rangle \quad |f(2\theta, E)|^2 \exp[-2M] \sum_{ij} \exp[i\mathbf{S} \cdot (\mathbf{r}_i - \mathbf{r}_j)], \tag{43}$$

given by the first term in Eq. (42). The second term in Eq. (42) is the one-phonon thermal diffuse scattering, and the third is the multiphonon scattering, which takes account of all higher-order phonon processes. The effect of thermal vibrations is to remove intensity from the zero-phonon diffraction maxima and distribute this intensity throughout the Brillouin zone as thermal diffuse scattering.

The above, although derived for penetrating radiation, is applicable to low-energy electron diffraction with some modifications (Strozier *et al.*, Volume I; Webb and Lagally, 1973). Slow electrons, because of their strong interaction with matter, are sensitive to only a limited number of atoms near the surface. LEED is therefore inherently a back-reflection technique, with consequently large momentum transfer **S**. The resulting large sensitivity of LEED intensities to thermal vibrations is illustrated in Fig. 17, which shows the zero-, one-, and multiphonon contributions to the intensity integrated over a Brillouin zone as a function of $2M = \langle (\mathbf{S} \cdot \mathbf{u})^2 \rangle$. Whereas for a rigid crystal $(2M = 0)$ all the intensity would be in the Bragg peak, at room temperature $2M$ is typically greater than 1 at energies of LEED experiments; thus less than one-third of the total intensity is actually in the zero-phonon scattering.

One feature of the large interaction of slow electrons with matter must be briefly considered. Whereas the interpretation of the thermal behavior of kinematic diffraction intensities in terms of the vibrational properties of

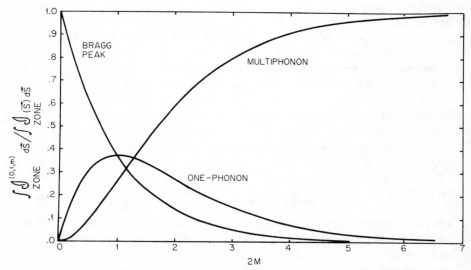

Fig. 17. The integrated intensity in a Brillouin zone in the zero-, one-, and multiphonon scattering as a function of $2M$.

the surface is relatively straightforward, this may not be true if multiple scattering is present. In Section B we will give qualitative arguments and some experimental demonstrations that multiple scattering does not cause significant errors in the determination of mean square amplitudes of vibration. However, since multiple scattering processes do make important contributions to the LEED intensity, efforts are usually made to analyze only those features that behave kinematically. Scattering from some surfaces appears to be mostly kinematic; in others kinematic features can usually be determined by observing the shape and motion of diffraction peaks as a function of diffraction geometry. The single scattering part can in many cases also be extracted from LEED intensity profiles by averaging several profiles at constant momentum transfer (Ngoc et al., 1973).

B. Surface Mean Square Vibrational Amplitudes

Surface mean square vibrational amplitudes are determined from the Debye–Waller factor for low-energy electrons, obtained by measuring the temperature dependence of the diffracted intensity. From Eq. (43),

$$\ln\left[I^{(0)}(T_1)/I^{(0)}(T_2)\right] = -2M_{eff} = -|S|^2 \langle u_s^2 \rangle_{eff}. \tag{44}$$

$\langle u_s^2 \rangle_{eff}$ is the effective mean square vibrational amplitude in the direction of S of the atoms participating in the scattering. Usually it is expressed as a ratio of the bulk value $\langle u^2 \rangle_{bulk}$ and sometimes as an effective Debye temperature,

$$\langle u_s^2 \rangle_{eff} \propto 1/\Theta_{D\,eff}^2. \tag{45}$$

An example of temperature dependence measurements for Ag (Dennis and Webb, 1973) is shown in Fig. 18. The curve shows three distinct regions: the low-temperature region where zero-point effects cause a deviation from linearity; the linear region where $\langle u^2 \rangle$ is proportional to temperature; and a region at high temperatures where the thermal diffuse scattering begins to dominate the temperature dependence and it is no longer possible to extract a $\langle u^2 \rangle$ without a correction for the phonon scattering.

Such measurements repeated at several energies will give the energy dependence of $\langle u_s^2 \rangle_{eff}$. This is as shown in Fig. 19 for the ratio $\langle u_\perp^2 \rangle_{eff}/\langle u_\perp^2 \rangle_{bulk}$ in Ni (Unertl and Webb, 1974). At all energies, the electron beam samples several atomic planes with different mean square vibrational amplitudes. As the energy increases, the electrons penetrate more deeply and see more planes with bulk-like vibrational amplitudes. Hence the measured effective mean square vibrational amplitude approaches the bulk value as the energy increases. Since the electron penetration is about the

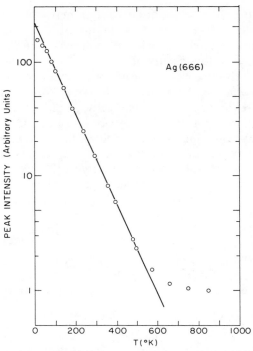

Fig. 18. Temperature dependence of the intensity of the (666) reflection in Ag(111). (After Dennis and Webb, 1973.)

same as the range in which $\langle u^2 \rangle$ is different from the bulk and changing rapidly, $\langle u^2 \rangle_{\text{surf}}$ cannot be extracted unless a model for the penetration is assumed. However, it is clear that $\langle u^2 \rangle_{\text{surf}}$ must be at least as large as the largest measured $\langle u^2 \rangle_{\text{eff}}$. Using the criterion

$$\langle u^2 \rangle_{\text{surf}} = (\langle u^2 \rangle_{\text{eff}})_{\text{max}}, \tag{46}$$

values for the normal component of the surface mean square vibrational amplitude range from about two to five times the bulk. These values can be interpreted only as a lower limit to the actual $\langle u^2 \rangle_{\text{surf}}$, since even at the energies where the largest values of $\langle u^2 \rangle_{\text{eff}}$ occur, more than one plane contributes to the scattering. To estimate how much larger $\langle u^2 \rangle_{\text{surf}}$ can be, a simple model for the penetration can be included in Eq. (41).

$$\langle I(\mathbf{S}) \rangle = |f(2\theta, E)|^2 \sum_{ij} |A_i| \, |A_j| \exp[i\mathbf{S} \cdot (\mathbf{r}_i - \mathbf{r}_j)] \, \langle \exp[i\mathbf{S} \cdot (\mathbf{u}_i - \mathbf{u}_j)] \rangle, \tag{47}$$

where A_i is the scattered amplitude contributed by the ith atom. Models for both A_i and $\mathbf{u}_i$ are required to fit measured intensities, since neither quantity is known by itself. Such fits have been made for only a few materials (Jones et al., 1966; Tong et al., 1973; Unertl and Webb, 1974; Buchholz, 1974). The A_i are determined either phenomenologically from the energy width of kinematic peaks in the intensity profile or from independent calculations. The vibrational amplitudes or their decay with depth into the crystal are taken from one of the model calculations of Section V. The resulting surface mean square vibrational amplitudes, listed in Table II, are larger, as expected, than those obtained by using Eq. (46).

In an analysis of Xe data, Tong et al. (1973) used calculated values of $\langle u_\perp^2 \rangle_n$ for three different models in an attempt to differentiate between them. Unfortunately, as already mentioned, such differentiation is not possible on the basis of the fit to the data. This is partly due to experimental uncertainty, but there may be two major reasons why anharmonic effects are difficult to distinguish. First, the effect of anharmonicity on the Debye–Waller factor tends to be obscured by the effect of thermal diffuse scattering. Anharmonicity should cause an increasingly steep slope in Fig. 18 as the temperature is raised; but at temperatures where anharmonicity becomes significant, $2M$ is large and the thermal diffuse scattering, which causes a flatter slope in Fig. 18 at higher temperatures, also becomes appreciable.

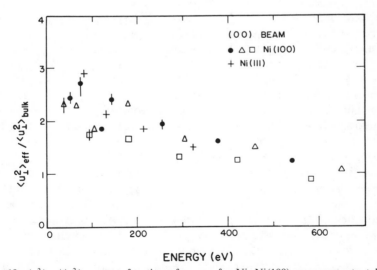

Fig. 19. $\langle u_\perp^2 \rangle_{\mathrm{eff}} / \langle u_\perp^2 \rangle_{\mathrm{bulk}}$ as a function of energy for Ni. Ni(100) measurements, taken at three different angles of incidence, show the dependence on penetration of the electron beam. ● $\vartheta = 29°$, △ $\vartheta = 19°$, □ $\vartheta = 7°$, where ϑ is measured from the surface normal. Error bars indicate experimental uncertainty. (After Unertl 1973.)

TABLE II

$$\langle u_\perp^2 \rangle_{\text{surf}} / \langle u_\perp^2 \rangle_{\text{bulk}}$$

Crystal	(a)	(b)	Model
1. Ag(111)	2.0	2.2 ± 0.4	Clark et al. (1965)
2. Ni(100)	2.5	4 ± 1	Clark et al. (1965)
3. Xe(111)	2 ± 0.5	3.48	anharmonic, Allen et al. (1969b)
		2.25	harmonic, Allen and de Wette (1969b)
		2.0	central force, Clark et al. (1965)

(a) assuming $\langle u^2 \rangle_{\text{surf}} = (\langle u^2 \rangle_{\text{eff}})_{\text{max}}$.
(b) Extracted from model calculations.
References: 1. Jones et al. (1966); 2. Unertl and Webb (1974); 3. Tong et al. (1973).

Second, the penetration is not well known. Apart from the uncertainty in the electron penetration at low energies where $\langle u_\perp^2 \rangle_{\text{eff}}$ is large, the amplitude contributed to the elastic scattering from the outer plane may actually be less than that from the second layer because of the larger vibrational amplitude of the outer layer. This temperature dependence of the penetration for elastic scattering, first pointed out by McKinney et al. (1967), results in a $\langle u_\perp^2 \rangle_{\text{eff}}$ that favors the deeper layers. It is partly responsible for the approach of $\langle u_\perp^2 \rangle_{\text{eff}}$ to bulk values at higher energies.

A large number of measurements of $\langle u_\perp^2 \rangle_{\text{eff}}$ for a variety of materials has been made in the high-temperature limit, as well as a few measurements of $\langle u_\parallel^2 \rangle_{\text{eff}}$ [for partial lists, see Somorjai and Farrell, (1972) and Tabor et al. (1971)]. The latter are much more difficult and hence more uncertain, but results in general are closer to bulk values. The analysis has usually rested upon Eq. (46) and thus gives surface vibrational amplitudes that are too low. Greater knowledge of the electron penetration is required before the uncertainty range indicated in Table II can be significantly reduced.

Recent measurements at very low temperatures have been able to detect the deviation from the high-temperature limit (Kaplan and Somorjai, 1971; Theeten et al., 1973; Dennis and Webb, 1973). By comparing the temperature dependence of $\langle u^2 \rangle_{\text{eff}}$ at low energies ($\simeq \langle u^2 \rangle_{\text{surf}}$) and at higher energies ($\simeq \langle u^2 \rangle_{\text{bulk}}$), Theeten et al. (1973) could detect differences that they suggest are due to differences in the force constants for surface and bulk atoms. LEED intensity measurements at low temperatures are, unfortunately, extremely sensitive to noble-gas adsorption, affecting the slope in Fig. 18 in the same manner as the zero-point vibrations. Hence great care must be taken to avoid such contamination.

It was mentioned earlier that multiple scattering may affect the determination of $\langle u^2 \rangle_{\text{eff}}$. If an electron is multiply scattered, each scattering contributes a term in the Debye–Waller factor (Somorjai and Farrell, 1972; Unertl, 1973) proportional to $\mathbf{S}_i$ for that scattering and $\langle u^2 \rangle_{\text{eff}}$ in the direction of $\mathbf{S}_i$. However, the attenuation of low-energy electrons is so strong that higher than second-order scatterings make a very small contribution to the elastically backscattered intensity. Second-order scatterings must consist of one forward and one backscattering; and since $\mathbf{S}$ for the forward scattering is small, its contribution to the Debye–Waller factor must also be small. Jona *et al.* (1972), in an exact multiple-scattering calculation for Ag(111), found Θ_{Deff} values, determined by assuming peaks in their intensity profiles were kinematic, differed by 8% from the values used as input into the calculation. Experimentally, most of the scatter is due to measurement technique (Gelatt *et al.*, 1970; Lagally, 1970), although pathological cases have been found (Reid, 1971; Unertl, 1973). $\langle u^2 \rangle_{\text{eff}}$ determined from individual profiles and averaged data are the same within experimental scatter (Ngoc *et al.*, 1973). Since for the great majority of situations $\langle u^2 \rangle_{\text{eff}}$ appears to be well behaved, one can conclude that multiple scattering does not affect the determination of mean square vibrational amplitudes from LEED intensities at the presently attainable level of accuracy.

C. Thermal Expansion

The manifestation of the anharmonicity of lattice vibrations in the bulk is thermal expansion; and since anharmonic effects are larger near the surface, one expects an excess surface thermal expansion. As for the bulk, the equilibrium atomic positions at any temperature can be obtained by minimizing the total free energy F with respect to the interatomic distances. At any temperature T, this gives the dynamic displacements, i.e., the sum of the static displacement plus the thermal expansion. In order to determine a thermal expansion coefficient for the surface, this procedure must be repeated over a range of temperatures. This has recently been done in the quasiharmonic approximation (Allen 1972; Dobrzynski and Maradudin, 1973; Kenner and Allen, 1973). Kenner and Allen, using the Lennard–Jones potential, find in the high-temperature limit

$$\frac{\alpha_{\text{surf}}}{\alpha_{\text{bulk}}} = \frac{3}{4} \frac{\langle u_\perp^2 \rangle_{\text{surf}}}{\langle u_\perp^2 \rangle_{\text{bulk}}}, \tag{48}$$

where α_{surf} is the total thermal expansion of the outer plane. Between its values at $T = 0°\,\text{K}$ and in the high-temperature limit, $\alpha_{\text{surf}}/\alpha_{\text{bulk}}$ has a large

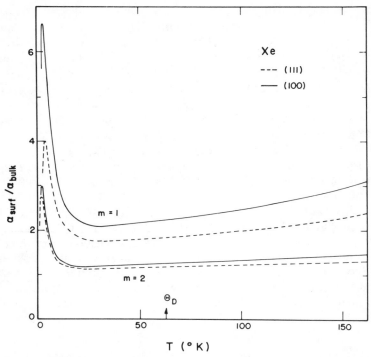

Fig. 20. Calculated temperature dependence of $\alpha_{\text{surf}}/\alpha_{\text{bulk}}$ for the outermost two planes of Xe. (After Kenner and Allen, 1973.)

peak as shown in Fig. 20. This can be explained in terms of the surface mode spectrum. Both α_{surf} and α_{bulk} rise like T^3 at low temperatures; but because of the presence of the lower-frequency surface modes, α_{surf}/T^3 peaks at a lower temperature than α_{bulk}/T^3. This gives rise to a peak in the ratio $\alpha_{\text{surf}}/\alpha_{\text{bulk}}$ at $\sim 6\%$ of the Debye temperature for this model. To what extent the simple high-temperature result is accurate if all anharmonic effects are accounted for remains to be tested. In any case, as seen in Section V, molecular dynamics calculations indicate both larger dynamic displacements and larger mean square vibrational amplitudes, implying that ratios $\alpha_{\text{surf}}/\alpha_{\text{bulk}}$ will actually be even larger than the value of ~ 2 found from Eq. (48).

Experimental evidence on the surface thermal expansion coefficient is in principle available from low-energy electron diffraction by measuring the energy shift of kinematic peaks in the intensity profile as a function of temperature. With increasing lattice constant, the peaks should shift to lower energies. The linear expansion coefficient is given by

$$\alpha = \frac{1}{d}\frac{\Delta d}{\Delta T}, \tag{49}$$

where d is the plane spacing. For single scattering, Bragg's law gives

$$\alpha_{\text{eff}} = -\frac{\Delta E}{2\Delta T(E + V_0)}, \tag{50}$$

where α_{eff} is the average thermal expansion coefficient of all planes contributing to the scattering, V_0 is the inner potential, and E is the energy of the electrons.

Only a few determinations of α_{eff} have been made (Gelatt et al., 1969; Woodruff and Seah, 1970; Wilson and Bastow, 1971; Reid, 1972; Ignatiev and Rhodin, 1973; Unertl and Webb, 1974). The general result on metals is that $\alpha_{\text{eff}} \simeq \alpha_{\text{bulk}}$, in contradistinction to the measurements of $\langle u^2 \rangle_{\text{eff}}$, which are always larger than $\langle u^2 \rangle_{\text{bulk}}$. That this is reasonable can be seen by the following argument (Unertl and Webb, 1974; Buchholz et al., 1975). Excess thermal expansion, just as excess vibrational amplitude, is in general limited to the outer plane. Whereas excess vibrational amplitude manifests itself in intensity changes, the expansion results in peak shape changes and shifts. These are under the best circumstances minor because the amplitude of only one plane out of the 5–10 involved in the diffraction is slightly out of phase at the peak positions. If in addition this plane does not contribute much amplitude, as is the case if it vibrates strongly, the effect on the diffracted-beam shapes and positions is minimal. It is paradoxical that in those systems where the thermal expansion is likely to be large (i.e., large surface vibrations), the expansion will not be detectable because the diffracted amplitude contributed by the affected plane is too small; and in those systems where the diffracted amplitude is large (i.e., small surface vibrations), the thermal expansion is likely to be too small to be detected. Hence only in unlikely pathological cases, where the surface thermal expansion is large but the surface vibrational amplitude is small, would one expect an α_{eff} different from α_{bulk}.

Any determination of α_{eff} must take into account the effect of multiple scattering, which also can cause peak shape changes and shifts. In the best experiments (Gelatt et al., 1969; Unertl and Webb, 1974), the peak shape was measured and did not change. This is a sufficient condition that the excess surface thermal expansion is too small to measure. However, the value of α_{eff} determined from the peak shifts still varied up to $\pm 40\%$ from α_{bulk}. Such uncertainties may be ascribable to multiple scattering or possibly very small experimental artifacts, but their magnitude points out the difficulty in measuring the excess surface thermal expansion.

VII. PHONON-ASSISTED SCATTERING OF LOW-ENERGY ELECTRONS

In the last section the effect of surface vibrations on the elastically scattered intensity was demonstrated. By measuring the decay of the intensity of diffraction maxima and their shift in energy, information about the mean square vibrational amplitude of surface atoms and about surface thermal expansion is obtained. However, since interaction with any phonon reduces the intensity of the elastic diffraction maximum, this gives no information about the phonon spectrum.

The intensity lost from the elastic peak by phonon-assisted scattering reappears throughout the Brillouin zone. It is easy to visualize this in the following manner. For a rigid lattice the periodic array of scatterers can be considered a density wave of wavelength a, the lattice constant, whose Fourier transform gives the diffracted beams. If now a second, long-wavelength, density fluctuation of given wavelength and direction is introduced into the solid (corresponding to a phonon of momentum $\hbar\mathbf{q}$, $|\mathbf{q}| = 2\pi/\lambda$) a sideband diffraction maximum will appear slightly displaced from the rigid-lattice peak in the direction of $\mathbf{q}$. The separation of this sideband from the rigid-lattice peak is proportional to the phonon momentum $\mathbf{q}$ and its height to the phonon mean square amplitude. The energy of electrons in the sideband differs from the elastic energy by the energy of the phonon $\hbar\omega_{\mathbf{q}}$. In principle the wavelength of this particular lattice wave can be measured by measuring the position of the sideband, and its frequency can be determined from a measurement of the energy loss or gain of the sideband electrons.

A. Thermal Diffuse Scattering

Since waves may travel in any direction in the lattice, with any wave-length down to $\lambda = 2a$, a continuous distribution of sidebands exists around each diffraction maximum. The intensity of this thermal diffuse scattering (TDS) depends on the energy of the electrons and the temperature, as well as the amplitude of different phonons. The TDS near the diffraction maxima is, of course, due to long-wavelength phonons and hence characteristic of an elastic continuum. Near the Brillouin zone boundary, dispersion effects become apparent. A significant portion of the TDS should be due to surface phonons because of the large part they play in the scattering of slow electrons.

One would like to extract the phonon dispersion relation from the TDS. To do so in a straightforward manner requires measuring the energy loss or gain $\Delta E = \hbar\omega_{\mathbf{q}}$ as well as the momentum $\mathbf{q}$ of a particular phonon at some point in the Brillouin zone. Neutron diffraction allows both of these measurements for individual phonons. In low-energy electron diffraction and x-ray diffraction systems, the angular resolution is adequate to measure the phonon momentum, but the energy resolution is insufficient to distinguish phonon-assisted diffraction from the truly elastic diffraction. The TDS is therefore often called "quasi-elastic." Because in three dimensions it is possible to choose directions in the crystal in which only one phonon can contribute at a particular value of $\mathbf{q}$, it is possible to use x-ray diffraction indirectly in some special cases to arrive at a dispersion relation (Wooster, 1962). At this value of $\mathbf{q}$, the intensity of the TDS is related directly to the mean square amplitude of the phonon, and from equipartition of energy it is possible to extract the phonon frequency. Because only the parallel component of momentum is conserved for scattering from surfaces, however, at any point $\mathbf{q}_{\parallel}$ in the Brillouin zone a range of phonons with different $q_{\perp}$ may contribute to the intensity (i.e., all phonons in the bands shown in Figs. 3 and 4 can contribute to the TDS). This is illustrated in terms of the scattering process in Fig. 21. It is thus not possible to determine the phonon dispersion relations with low-energy electron diffraction, and the TDS must be calculated from models of the phonon spectrum. The simplest such model (McKinney *et al.*, 1967), considers the kinematic scattering of nonpenetrating radiation from the outer plane of a lattice whose thermal motion is described by a Debye spectrum in the high-temperature limit. McKinney *et al.* evaluated the one-phonon scattering [the second term in

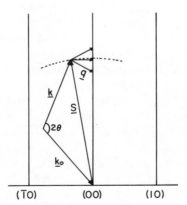

(TO) (OO) (IO)

Fig. 21. Schematic of thermal diffuse scattering for nonpenetrating radiation. All phonons with given $\mathbf{q}_{\parallel}$ can contribute to the scattering at S. The dotted line indicates the path of the detector for measurement of the distribution of TDS in the Brillouin zone.

Eq. (42)] and compared it with measurements of the angular shape of
diffraction maxima from Ag(111). The effects of the finite penetration of the
beam and the phonon boundary conditions on the model were also
investigated. This model predicts the correct energy, temperature, and $|\mathbf{q}_{\parallel}|$
dependence of the TDS. Similar calculations for an isotropic elastic
continuum have been made by Huber (1967), who finds that surface modes
account for about half the TDS near the elastic maxima. This comes about
because of the much larger amplitude of the surface modes, since the energy
of a surface mode is shared among fewer atoms.

More detailed calculations of the one-phonon thermal diffuse scattering
have recently been made with a lattice-dynamics model in the quasiharmonic
approximation (de Wette and Allen, 1969; Kesmodel *et al.*, 1972, 1973).
Kinematic slow-electron scattering from Xe is considered, with the electron
penetration taken into account. The model clearly illustrates the effect of the
finite penetration of the beam on the shape and distribution of the TDS.
For example, scattering from a single Xe(111) layer at any $S_{\perp}$ gives a
sixfold symmetric distribution of TDS around the reciprocal-lattice rods.
This anisotropy is due to the anisotropy in the velocity of sound; for an
isotropic medium, no anisotropy will be found. If more than one plane
scatters, the distribution around different reciprocal-lattice rods will be
different and dependent on the attenuation of the beam. No measurements
of the thermal diffuse scattering on noble-gas solids exist, but comparisons
with measurements on Ag(111) (Dennis and Webb, 1973) show good
agreement in the central part of the Brillouin zone.

In addition to the one-phonon scattering, there is a multiphonon contribution, given by the third term in Eq. (42), due to scatterings in which
two or more phonons participate. The multiphonon scattering (Barnes *et al.*,
1968) causes a uniform background intensity throughout the Brillouin zone
and becomes the dominant form of scattering at high temperatures (see Fig.
17). It is the main cause for the deviation from linearity in the slope of
Debye–Waller plots at high temperatures illustrated in Fig. 18. Its properties
are useful in determining the effective scattering factor for atoms in the
crystal (Lagally and Webb, 1969), but it complicates analysis of the TDS.
It is especially serious near the Brillouin zone boundary, where the one-
phonon scattering is small. But since the scattering here is sensitive to short-
wavelength phonons, the most information about surface vibrations is
available near the zone boundary. A detailed separation of the zero-, one-,
and multiphonon scattering from Ag(111) has recently been made by Dennis
and Webb (1973) in order to isolate the one-phonon scattering. Figure 22
shows the three components. The phonon dispersion is clearly evident in
the one-phonon scattering. In order to determine the dispersion relations

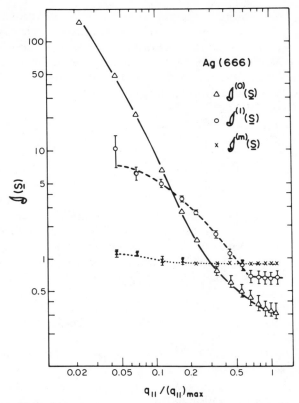

Fig. 22. Separation of the zero-, one-, and multiphonon components of the scattered intensity. Measurements are made at $S_{\perp}$ corresponding to the (666) reflection from Ag(111). (After Dennis and Webb, 1973.)

for the surface phonons, realistic calculations for different models of the phonon spectrum must be compared to such measurements. These calculations do not as yet exist for metals.

B. Inelastic Electron Spectroscopy

If, as in neutron scattering, the energy change in the scattered beam due to the interaction with a phonon is resolvable, the frequency of the phonon can be measured directly. Since phonon energies are typically of the order of several tens of meV, this requires a specially designed high-resolution detector as well as a highly monochromatic electron source. Such a system has been used recently by Ibach (1970) for investigations of optically active surface phonons. The thermal diffuse scattering discussed in the last section can be called "impact scattering," the scattering of electrons off the vibrating localized atomic potentials of the lattice atoms. For optically active crystals there may also be "dipole scattering," the interaction of electrons with the dipole field of certain optical phonons, as discussed in Section III.C.

For a continuum with dielectric constant $\varepsilon(\omega) = \varepsilon_1 + i\varepsilon_2$, the energy loss probability in the vicinity of a boundary surface is given by (Sevier, 1972)

$$W \propto \text{Im}\left[\frac{1}{\varepsilon(\omega)+1}\right], \qquad (51)$$

which has a resonance at $\varepsilon_1 = -1$, unlike the volume loss probability, which has a maximum at $\varepsilon_1 = 0$. The latter corresponds to excitation of longitudinal bulk phonons, which will not be excited in the former. Hence for reflection of electrons that do not penetrate into the bulk, only optical surface phonons will be excited. The electron interaction with the dipole field of the phonon will have a resonance when the parallel component of the electron velocity is equal to the phase velocity of the phonon, $\mathbf{v}_\parallel = \omega_s/\mathbf{q}_\parallel$. Since for all practical electron energies $E \gg \hbar\omega_s$, the electron interacts chiefly with phonons of small $|\mathbf{q}_\parallel|$, i.e., the scattering is concentrated in the vicinity of the elastic peak. The loss probability increases with decreasing energy.

Measurements on ZnO (Ibach, 1970) reveal a series of equally spaced loss peaks in the direction of the diffracted beam, as well as a much smaller energy gain peak whose magnitude is related to the first loss by the Boltzmann factor. This is illustrated in Fig. 23. The higher-order loss peaks are interpreted as multiphonon processes. The loss energies agree very well with ones calculated for the surface phonon from experimental values of $\varepsilon(\omega)$.

A similar loss peak is observed in Si(111) (Ibach, 1971b). In bulk Si the optical phonons have no associated electric fields and are therefore not optically active. However, the reduced symmetry and change in geometry of atoms near the surface may cause a dipole field to be set up by the vibration of two sublattices against each other. Hence it is possible that the existence of a surface may lead to a distinct class of phonons that is absent in the bulk.

In principle, a spectrometer with good energy and angular resolution should permit a mapping of the impact scattering as well, by measuring the range of energy losses for electrons at different $\mathbf{q}_\parallel$ in the Brillouin zone. Observed spectra at a particular $\mathbf{q}_\parallel$ should then show (Roundy and Mills,

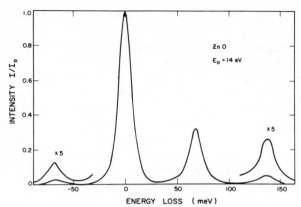

Fig. 23. Energy loss spectrum for 14 eV electrons scattered from ZnO (1100). (After Ibach, 1971.)

1972) a band of frequencies corresponding to the allowed range of $q_\perp$ for each $\mathbf{q}_\parallel$ (the bulk bands), as well as discrete frequencies for surface modes lying outside the bulk bands.

VIII. ATOMIC-BEAM SCATTERING

Slow-electron scattering is the only technique that has so far significantly increased our understanding of surface vibrations. However, the scattering of atomic or molecular beams holds some promise for measuring the mean square vibrational amplitude of surface atoms and possibly the energy loss to particular surface phonons. Because the scattering is from the outer plane, multiple scattering is not important, and analysis of measurements in terms of properties of the surface should be in principle simpler. In practice, the results are greatly dependent on the form and strength of the incident atom–surface interaction potential, which is at present not well known.

A description of the scattering process is given by the "hard-cube" (Logan and Stickney, 1966) or "soft-cube" (Logan and Keck, 1968) models. The models differ in the form of the atom–surface interaction potential; the soft-cube model, by considering a stationary attractive plus exponential repulsive potential, allows for refraction of the atom beam, which is not possible with the purely impulsive repulsive and zero attractive potential of the hard-cube model. Both models assume that the interaction potential is uniform in the plane of the surface, i.e., that the surface is perfectly smooth; and therefore the parallel component of the momentum of the beam is conserved.

The elastic scattering of particles by these potentials is easily visualized. For scattering at low temperatures a specular beam with a narrow, well-defined angular range should be observed. At higher temperatures the momentum transfer normal to the surface $S_\perp$ is either larger or smaller than for the elastic scattering because of the motion of the cube. This leads to a broadened elastic peak. These simple models, however, cannot account for many-body effects like phonon scattering of the atom, which require that cubes interact with each other and that the atom interacts with many cubes during its collision time. Proceeding from the basis that such interactions exist, an analysis of the height and angular shape of the peak and energy loss of scattered particles provides the possibility for studying surface lattice dynamics.

Beeby (1971) has shown that in principle the normal component of the mean square vibrational amplitude at the surface can be determined from a decrease of the elastic peak intensity with increasing temperature. As in the kinematic scattering of other radiation, the thermal motion can be thought of as smearing out the potential with a consequent reduction of the elastically scattered intensity. However, because the attractive potential is weak and slowly varying, it is effectively time-independent and only the repulsive part is smeared. The intensity can then be written proportional to a Debye–Waller factor

$$I \propto \exp[-S_\perp^2 \langle u_\perp^2 \rangle], \tag{52}$$

where $S_\perp$ is the momentum transfer evaluated inside the attractive potential, which is much larger than $S_\perp$ outside. Unfortunately, because the shape and depth of the atom–surface interaction potential are not well known, $S_\perp$ can not be evaluated without a model. Evaluation of data for scattering of He or H_2 from several surfaces (Weinberg, 1972) using a simple one-conduction-electron-per-atom model to determine the interaction potential depth (Beeby, 1971) gives $\langle u_\perp^2 \rangle$ values that differ widely from those obtained from LEED and are in some cases smaller than the accepted bulk $\langle u_\perp^2 \rangle$ values.

Elastic-peak intensity measurements can, of course, give information only about the mean square vibrational surface amplitude. To learn about the phonon spectrum requires analysis of the angular shape of the scattered beam. Although considerable work of a qualitative nature has been done, as yet no clear pattern of consistency emerges when peak widths are compared to lattice-dynamical quantities (Weinberg and Merrill, 1972).

Finally, additional information about the vibrational modes of the surface may be gained by comparison of atomic and molecular scattering. Whereas the translational modes of the impinging gas atoms have quite different energy than the surface vibrational modes, making energy transfer difficult, rotational or vibrational modes of a molecule may be much nearer in energy. The broad angular distributions obtained for molecular scattering (Palmer *et al.*, 1969) are interpreted as arising from the stronger inelastic scattering due to rotation–vibration energy transfer. Energy analysis of the scattered beam has not been attempted so far.

At present, it appears that the interaction potential is not well enough known to use atomic scattering for quantitative determinations of surface vibrational amplitudes. Rather, it seems more probable that LEED results will be used in gas–surface interactions to determine the interaction potential. Once this is known, however, atomic beam scattering is likely to become an important tool in the study of surface vibrations.

IX. CONCLUSIONS

This chapter has attempted to cover the major aspects of thermal vibrations of clean surfaces. The introduction of a surface, because it behaves like an extended defect in the solid, causes changes in the normal mode spectrum and hence in a variety of quantities that depend upon it. It seems fair to say that calculational methods have reached a stage of sophistication that allows inclusion of all anharmonic effects, at least for simple potentials. Advances can be expected in realistically modeling metals in the near future. So far sufficiently detailed and extensive experimental information is available only on the mean square vibrational amplitudes of metals in the high-temperature limit. Studies of surface phonons are still quite few. Considerable activity is likely as experimental methods become more sophisticated. However, the real future of studies of surface vibrations appears to lie in other directions, such as the more complex chemisorption or physisorption systems where knowledge of the mean square vibrational amplitude or the frequency of localized modes will provide information on the binding of the adsorbed atom, or in thin films or layered structures where increased knowledge of the effect of surface vibrations may lead to new devices of technological importance.

ACKNOWLEDGMENTS

The author would like to thank M. B. Webb and J. C. Buchholz for reading the manuscript. The preparation of this chapter was supported in part by the Research Corporation and the National Science Foundation.

REFERENCES

ALDER, B. J., VAISNYS, J. R., and JURA, G. (1959). *J. Phys. Chem. Solids* **11**, 182.
ALLEN, R. E. (1972). *J. Vac. Sci. Technol.* **9**, 934.
ALLEN, R. E., and DE WETTE, F. W. (1969a). *Phys. Rev.* **179**, 873.
ALLEN, R. E., and DE WETTE, F. W. (1969b). *J. Chem. Phys.* **51**, 4820.
ALLEN, R. E., and DE WETTE, F. W. (1969c). *Phys. Rev.* **188**, 1320.
ALLEN, R. E., ALLDREDGE, G. P., and DE WETTE, F. W. (1969a). *Phys. Rev. Lett.* **23**, 1285.
ALLEN, R. E., DE WETTE, F. W., and RAHMAN, A. (1969b). *Phys. Rev.* **179**, 887.
ALLEN, R. E., ALLDREDGE, G. P., and DE WETTE, F. W. (1971a). *Phys. Rev.* **B4**, 1648.
ALLEN, R. E., ALLDREDGE, G. P., and DE WETTE, F. W. (1971b). *Phys. Rev.* **B4**, 1661.
BARKER, A. S., Jr. (1972). *Phys. Rev. Lett.* **28**, 892.
BARNES, R. F., LAGALLY, M. G., and WEBB, M. B. (1968). *Phys. Rev.* **171**, 627.
BEEBY, J. L. (1971). *J. Phys.* **C4**, L353.
BENSON, G. C., and YUN, K. S. (1967). *In* "The Solid-Gas Interface" (E. A. Flood, ed.) Dekker, New York.
BOERSCH, H., GEIGER, J., and STICKEL, W. (1968). *Z. Phys.* **212**, 130.
BORN, M., and HUANG, K. (1954). "Dynamical Theory of Crystal Lattices." Oxford Univ. Press, Clarendon (1954).
BRYKSIN, V. V., GERBSHTEIN, YU. M., and MIRLIN, D. N. (1972). *Sov. Phys. Solid State* **13**, 1779.
BUCHHOLZ, J. C., (1974). Ph.D. Dissertation. Univ. of Wisconsin (unpublished).
BUCHHOLZ, J. C., WANG, G.-C., and LAGALLY, M. G. (1975). *Surface Sci.* **49**, (to be published).
BURRIDGE, E. (1970). *Quart. J. Mech. Appl. Math.* **23**, 217.
BURTON, J. J., and JURA, G. (1967). *J. Phys. Chem.* **21**, 1932.
CHASE, J. B., and KLIEWER, K. L. (1970). *Phys. Rev.* **B2**, 4389.
CHEN, T. S., ALLEN, R. E., ALLDREDGE, G. P., and DE WETTE, F. W. (1970). *Solid State Commun.* **8**, 2105.
CHEN, T. S., ALLDREDGE, G. P., DE WETTE, F. W., and ALLEN, R. E. (1971a). *Phys. Rev. Lett.* **26**, 1543.
CHEN, T. S., ALLDREDGE, G. P., DE WETTE, F. W., and ALLEN, R. E. (1971b). *J. Chem. Phys.* **55**, 3121.
CHEN, T. S., ALLDREDGE, G. P., DE WETTE, F. W., and ALLEN, R. E. (1972a). *Phys. Rev.* **B6**, 623.
CHEN, T. S., ALLDREDGE, G. P., and DE WETTE, F. W. (1972b). *Phys. Lett.* **40A**, 401.
CHEN, T. S., ALLDREDGE, G. P., and DE WETTE, F. W. (1972c). *Solid State Commun.* **10**, 941.
CHENG, D. J., WALLIS, R. F., and DOBRZYNSKI, L. (1974). *Surface Sci.* **34**, 400.
CLARK, B. C., HERMAN, R., and WALLIS, R. F. (1965). *Phys. Rev.* **139**, A860.
DENNIS, R. L., and HUBER, D. L. (1972). *Phys. Rev.* **B5**, 4717.
DENNIS, R. L., and WEBB, M. B. (1973). *J. Vac. Sci. Technol.* **10**, 192.
DE WETTE, F. W., and ALLEN, R. E. (1969). *In* "The Structure and Chemistry of Solid Surfaces" (G. A. Somorjai, ed.). Wiley, New York.
DOBRZYNSKI, L., and MARADUDIN, A. A. (1973). *Phys. Rev.* **B7**, 1207.
ENGLEMAN, R., and RUPPIN, R. (1966). *Phys. Rev. Lett.* **16**, 898.
ENGLEMAN, R., and RUPPIN, R. (1968a). *J. Phys. C* **1**, 614, 630, 1515.
ENGLEMAN, R., and RUPPIN, R. (1968b). *In* "Localized Excitations in Solids" (R. F. Wallis, ed.). Plenum Press, New York.
EVANS, D. J., USHIODA, S., and MCMULLEN, J. D. (1973). *Phys. Rev. Lett.* **31**, 369.
FUCHS, R., and KLIEWER, K.L. (1965). *Phys. Rev.* **140**, A2076.
FUCHS, R., KLIEWER, K. L., and PARDEE, W. J. (1966). *Phys. Rev.* **150**, 589.
GAZIS, D. C., HERMAN, R., and WALLIS, R. F. (1960). *Phys. Rev.* **119**, 533.
GELATT, C. D., Jr., LAGALLY, M. G., and WEBB, M. B. (1969). *Bull. Amer. Phys. Soc.* **14**, 793.

GELATT, C. D., Jr., LAGALLY, M. G., and WEBB, M. B. (1970). *Bull. Amer. Phys. Soc.* **15**, 632.
GOLDANSKII, V. I., and SUZDALEV, I. P. (1971). *In* Proc. Conf. Appl. Mössbauer Effect (I. Dézi, ed.) Akad, Kiadò, Budapest.
HUBER, D. L. (1967). *Phys. Rev.* **153**, 772.
IBACH, H. (1970). *Phys. Rev. Lett.* **24**, 1416.
IBACH, H. (1971a). *Festkörperprobleme* **11**, 135.
IBACH, H. (1971b). *Phys. Rev. Lett.* **27**, 253.
IGNATIEV, A. and RHODIN, T. N. (1973). *Phys. Rev.* **B8**, 893.
JACKSON, D. P. (1971). *Can. J. Phys.* **49**, 2093.
JACKSON, D. P. (1974). *Surface Sci.* **43**, 431.
JAMES, R. W. (1965). "The Optical Principles of the Diffraction of X-rays." Bell and Sons, London.
JEPSEN, D. W., MARCUS, P. M., and JONA, F. (1972). *Phys. Rev.* **B6**, 3684.
JONA, F., JEPSEN, D. W., and MARCUS, P. M. (1972). *In* Proc. LEED Seminar, 6th, Washington. (unpublished).
JONES, W. E., and FUCHS, R. (1971). *Phys. Rev.* **B4**, 3581.
JONES, E. R., Jr., McKINNEY, J. T., and WEBB, M. B. (1966). *Phys. Rev.* **151**, 476.
KAPLAN, R., and SOMORJAI, G. A. (1971). *Solid State Commun.* **9**, 505.
KELLERMANN, E. W. (1940). *Phil. Trans. Roy. Soc. London* **A238**, 513.
KENNER, V. E., and ALLEN, R. E. (1973). *Phys. Rev.* **B8**, 2916.
KESMODEL, L. L., DE WETTE. F. W., and ALLEN, R. E. (1972). *Solid State Commun.* **11**, 145.
KESMODEL, L. L., DE WETTE, F. W., and ALLEN, R. E. (1973). *Phys. Rev.* **B7**, 802.
KLIEWER, K. L., and FUCHS, R. (1966a). *Phys. Rev.* **144**, 495.
KLIEWER, K. L., and FUCHS, R. (1966b). *Phys. Rev.* **150**, 573.
LAGALLY, M. G. (1970). *Z. Naturforsch.* **25a**, 1567.
LAGALLY, M. G., and WEBB, M. B. (1969). *In* "The Structure and Chemistry of Solid Surfaces" (G. A. Somorjai, ed.). Wiley, New York.
LAGALLY, M. G., NGOC, T. C., and WEBB, M. B. (1971). *Phys. Rev. Lett.* **26**, 1557.
LARAMORE, G. E. (1973). *Phys. Rev.* **B8**, 515.
LIM, T. C., and FARNELL, G. W. (1968). *J. Appl. Phys.* **39**, 4319.
LIM, T. C., and FARNELL, G. W. (1969). *J. Acoust. Soc. Amer.* **45**, 845.
LOGAN, R. M., and KECK, J. C. (1968). *J. Chem. Phys.* **49**, 860.
LOGAN, R. M., and STICKNEY, R. F. (1966). *J. Chem. Phys.* **44**, 195.
LUCAS, A. A. (1968). *J. Chem. Phys.* **48**, 3156.
LÜTH, H. (1972). *Phys. Rev. Lett.* **29**, 1377.
MACRAE, A. U. (1964). *Surface Sci.* **2**, 522.
MARADUDIN, A. A. (1966). *Solid State Phys.* **19**, 2.
MARADUDIN, A. A., MONTROLL, E. W., WEISS, G. H., and IPATOVA, I. P. (1971). "Theory of Lattice Dynamics in the Harmonic Approximation." Academic Press, New York.
MARSCHALL, N., and FISCHER, B. (1972). *Phys. Rev. Letts.* **28**, 811.
MARTIN, M. R., and SOMORJAI, G. A. (1973). *Phys. Rev.* **B7**, 3607.
McKINNEY, J. T., JONES, E. R., and WEBB, M. B. (1967). *Phys. Rev.* **167**, 523.
MENZEL–KOPP, C., and MENZEL, E. (1956). *Z. Phys.* **144**, 538.
MORRISON, J. A., and PATTERSON, D. (1956). *Trans. Faraday Soc.* **52**, 764.
MUSSER, S. W., and RIEDER, K. H. (1970). *Phys. Rev.* **B2**, 3034.
NAUGLE, D. G., BAKER, J. W., and ALLEN, R. E. (1973). *Phys. Rev.* **B7**, 3028.
NGOC, TRAN C., LAGALLY, M. G., and WEBB, M. B. (1973). *Surface Sci.* **35**, 117.
NOVOTNY, V., and MEINCKE, P. P. M. (1973). *Phys. Rev.* **B9**, 4186.
PALMER, R. L., SALTSBURG, H., and SMITH, J. N. Jr. (1969). *J. Chem. Phys.* **50**, 4661.
LORD RAYLEIGH (1885). *London Math. Soc. Proc.* **17**, 4.
REID, R. (1971). *Phys. Status Solidi* **4a**, K211.

REID, R. (1972). *Surface Sci.* **29**, 623.
RIEDER, K. H., and HÖRL, E. M. (1968). *Phys. Rev. Lett.* **20**, 209.
ROUNDY, V., and MILLS, D. L. (1972). *J. Vac. Sci. Technol.* **9**, 699.
RUPPIN, R., and ENGLEMAN, R. (1970). *Rep. Progr. Phys.* **33**, 149.
SEVIER, K. D. (1972). "Low-Energy Electron Spectrometry." Wiley (Interscience), New York.
SOMORJAI, G. A., and FARRELL, H. H. (1972). *Advan. Chem. Phys.* **20**, 215.
STONELEY, R. (1955). *Proc. Roy. Soc. (London)* **A232**, 447.
STRATTON, R. (1953). *Phil. Mag.* **44**, 519.
TABOR, D., WILSON, J. M., and BASTOW, T. J. (1971). *Surface Sci.* **26**, 471.
THEETEN, J. B., and DOBRZYNSKI, L. (1972). *Phys. Rev.* **B5**, 1529.
THEETEN, J. B., DOMANGE, J. L., and HURAULT, J. P. (1973). *Surface Sci.* **35**, 145.
TONG, S. Y., and MARADUDIN, A. A. (1969). *Phys. Rev.* **181**, 1318.
TONG, S. Y., RHODIN, T. N., and IGNATIEV, A. (1973). *Phys. Rev.* **B8**, 906.
UNERTL, W. N. (1973). Ph.D. Dissertation, Univ. of Wisconsin (unpublished).
UNERTL, W. N., and WEBB, M. B. (1974). *J. Vac. Sci. Technol.* **11**, 193.
VAIL, J. M. (1967). *Can. J. Phys.* **45**, 2661.
WALLIS, R. F., MILLS, D. L., and MARADUDIN, A. A. (1968). *In* "Localized Excitations in Solids" (R. F. Wallis, ed.). Plenum Press, New York.
WALLIS, R. F., CLARK, B. C., and HERMAN, R. (1968). *Phys. Rev.* **167**, 652.
WEBB, M. B., and LAGALLY, M. G. (1973). *Solid State Phys.* **28**, 301.
WEINBERG, W. H. (1972). *J. Chem. Phys.* **57**, 5463.
WEINBERG, W. H., and MERRILL, R. P. (1972). *J. Chem. Phys.* **56**, 2881.
WILSON, J. M., and BASTOW, T. J. (1971). *Surface Sci.* **26**, 461.
WOLFRAM, T., DE WAMES, R. E., and KRAUT, E. A. (1972). *J. Vac. Sci. Technol.* **9**, 685.
WOODRUFF, D. P., and SEAH, M. P. (1970). *Phys. Status Solidi* **A1**, 429.
WOOSTER, W. A. (1962). "Diffuse X-ray Reflections from Crystals." Oxford Univ. Press (Clarendon), London and New York.

SUPPLEMENTARY REFERENCES

The following reviews cover particular aspects of this article. If cited earlier, only the names are given.

BENSON, G. C., and YUN, K. S. (1967).
DRANSFELD, K., and SALZMANN, E. (1970). *Phys. Acoust.* **7**, 219.
DUKE, C. B. (1972). *In* "LEED-Surface Structure of Solids" (M. Láznička, ed.). Union of Czechoslovak Mathematicians and Physicists, Prague.
GOLDANSKII, V. I., and SUZDALEV, I. P. (1971).
IBACH, H. (1971a).
IBACH, H. (1972). *J. Vac. Sci. Technol.* **9**, 713.
LAGALLY, M. G. (1972). *In* "LEED-Surface Structure of Solids." (M. Láznička, ed.), Union of Czechoslovak Mathematicians and Physicists, Prague.
LUDWIG, W. (1967). *Springer Tracts Mod. Phys.* **43**, 1.
MARADUDIN, A. A., MONTROLL, F. W., WEISS, G. H., and IPATOVA, I. P. (1971).
MARADUDIN, A. A. (1966).
RUPPIN, R. and ENGLEMAN, R. (1970).

SOMORJAI, G. A., and FARRELL, H. H. (1972).
STICKNEY, R. E. (1967). *Advan. At. Mol. Phys.* **3**, 143.
WALLIS, R. F. (1972). *In* "Atomic Structure and Properties of Solids" (E. Burstein, ed.). Academic Press, New York.
WOLFRAM, T., DE WAMES, R. E., and KRAUT, E. A. (1972).
WEBB, M. B., and LAGALLY, M. G. (1973).

10

Interaction between Surfaces: Adhesion and Friction

D. TABOR

DEPARTMENT OF PHYSICS
CAVENDISH LABORATORY, UNIVERSITY OF CAMBRIDGE
CAMBRIDGE, ENGLAND

I. INTRODUCTION

This survey is concerned with the adhesion and friction between solid surfaces; it attempts, where possible, to explain observed phenomena in terms of atomic processes or, failing that, in terms of bulk properties. In dealing with adhesion little reference will be made to adhesives, since in the last decade a number of extremely good monographs have appeared dealing with this theme (Eley, 1962; Weiss, 1962; Houwink and Salomon, 1967; Patrick, 1967; Bickermann, 1968).[†]

Adhesion and friction both involve contact between solid surfaces. Therefore we shall first discuss the nature and topography of solid surfaces and what happens when two such surfaces are placed in contact and/or pressed together. We shall then discuss the types of forces operating over the regions of atomic or near-atomic contact and the role they play in adhesion and friction.

II. THE NATURE AND TOPOGRAPHY OF SOLID SURFACES: CONTACT

A. The Nature of Solid Surfaces

There are many techniques for studying solid surfaces. These have been discussed in previous chapters of this book, but the main conclusions relevant to this chapter are as follows. The structure is best studied with low-energy electron diffraction (LEED) at normal incidence and high-energy

---

[†] Soon after this survey was completed, Dr. L. H. Lee kindly presented me with a copy of "Recent Advances in Adhesion" (Gordon and Breach, 1973). I must apologize to authors of many of the important contributions appearing in that volume for not quoting their papers in this survey.

electron diffraction (HEED) at glancing incidence or by field ion microscopy, though this necessitates the formation of fine tips of very small radius of curvature. All these studies show how difficult it is to obtain pure clean surfaces. The most elaborate steps, involving heating in very high vacuum, ion bombardment and preferential chemical attack, may all be needed to remove the last traces of surface contaminant. Even so, small quantities of bulk impurity (of order 1 ppm) may diffuse to the surface during the cleaning process. When finally cleaned, a surface can be kept clean in a high vacuum for fairly protracted periods. A safe generalization is that at 10^{-6} Torr a monolayer of gas will take about 1 sec to form, assuming a sticking coefficient of unity, and at 10^{-10} Torr, about 10^4 sec, i.e., over 2 hours.

B. Solid Surfaces in Air

Metal surfaces are of primary interest because of their wide use in engineering practice. Even if they are noble, they immediately adsorb oxygen and water vapor if exposed to the air; this film may not be more than a few molecules thick (Bowden and Throssel, 1951). With reactive metals there is initially some chemisorption that soon gives way to chemical reactions and the formation of oxides or hydroxides. During the early stages of oxide growth there may be rearrangement of the surface atoms of the metal that, in a more exaggerated form, can lead to faceting. The oxide may then grow to a thickness of the order of 50–100 Å by a process of diffusion of metal ions through the oxide to the oxide–gaseous interface. Some oxides are epitaxial and firmly linked to the substrate; they can impede diffusion. In that case the oxidation soon comes to an end, and the material appears to be noncorrosive or even "noble" (e.g., chromium, rhodium). In other cases the oxide is in marked mismatch with the substrate and cracks as it grows, so exposing fresh metal for further oxidation. Oxides may be amorphous or crystalline; they may be smooth or they may grow as crystals and whiskers so that the surface resembles a disorderly array of skyscrapers (Pfefferkorn, 1953; Takagi, 1957).

Ionic solids and covalent solids are usually materials of high surface energy. However, in the atmosphere adsorption rarely leads to the presence of more than a few molecular layers of oxygen or water vapor. On the other hand, the first monolayer may be extremely strongly bound to the surface. In some cases the interaction may involve an appreciable change in the surface. For example, water vapor appears to plasticize the surface of glass and of a *polar* polymer such as nylon. At higher temperatures oxygen can interact with diamond to facilitate graphitization and, at a later stage, combustion. Indeed at high temperatures oxygen will react chemically with the surface of many ionic and covalent materials.

Van de Waals solids have low surface energy, are usually hydrophobic, and are relatively inert. Although there have been a number of studies of ideal van der Waals solids such as solid argon and xenon, these are not engineering materials and are of little interest in the world of practical affairs. On the other hand rubber and polymers are of increasing importance. Generally these materials are not single substances but contain chemical homologues, plasticizers, and other additives; in addition they are only partially crystalline, or completely amorphous. Although relatively inert, they tend to oxidized slowly in the atmosphere, particularly in the presence of light, and the surface layers tend to crack.

C. The Topography of Solid Surfaces

There are several texts that discuss methods of studying the topgraphy of surfaces. The simplest method is by optical microscopy, especially if a cross section or an oblique cross section of the surface is prepared. Profilometry gives much higher vertical resolution (of the order of 100 Å, while in the more recent models of "Talystep" the vertical resolution is as low as 10 Å), but the lateral resolution is poor and not better than about 1 μm. Both techniques give only a line section. However Williamson (1968) has described an extension of the profilometer technique in which a series of closely spaced parallel profiles is recorded; from this a "two-dimensional contour" can be derived. The method that he describes as microcartography demands extremely accurate instrumentation. By contrast optical interference between the surface and a reference optical flat provides a relatively simple means of deriving surface contours. Here the vertical resolution for two-beam interference is of the order of a few hundred angstroms, but the lateral resolution cannot be better than a few thousand. The interferograms do, however, provide a very realistic two-dimensional picture of the surface. Another technique now receiving wide application is the scanning electron microscope. The surfaces can be studied directly without the use of replicas, though if the surface in noconducting a thin conducting film must be evaporated on to it. The depth of focus is remarkable, and features of the order of a few hundred angstroms in height and width can be resolved. The best resolution (down to 10 Å or less) is obtained in transmission electron microscopy, though a surface replica is generally needed. An original technique, the "grafiner," has recently been developed for surface contour studies (Young, 1972), but it is not yet available commercially.

All these techniques show that most surfaces are not atomically smooth over more than very restricted areas. Cleared surfaces usually show cleavage steps, electropolished surfaces undulations, and evaporated films orange-peel texture. Even surfaces prepared by solidifying the liquid usually contain roughnesses, fissures, or irregular growth features. The only material that appears to provide relatively large molecularly smooth areas is mica; this can be cleaved along a single molecular plane over areas of several square centimeters.

Some studies of adhesion and friction have been carried out on very carefully prepared surfaces in high vacuum where the surfaces are probably very smooth (for example, the experiments of Buckley, 1969), but most studies have been made on engineering-type surfaces where the roughness is relatively coarse. With the majority of solids mechanical methods of surface preparation such as cutting, grinding, abrasion, or polishing generate roughnesses that are very coarse on an atomic scale. In addition they produce distortions and modifications in the outermost layers. The resulting surface may be very complex indeed. For example, in the polishing of metals the surface layers may consist of a smeared "fudge" of metal, metal oxide, and polishing powder. On top of this layer a new oxide forms that will have a structure and topography depending on the conditions of growth. A diagrammatic representation of this is shown in Fig. 1. Clearly such surfaces are neither smooth nor simple.

Most of the systematic studies of surface topography have been carried out on mechanically prepared surfaces. This is partly because these are the techniques most commonly used in engineering practice. In addition, in friction and adhesion studies one of the easiest ways of removing grease and similar contaminant films is by abrasion under running water. Consequently this section will have little to say of the natural surface of glass or the cleavage surfaces of rock-salt or diamond. However, it is probable that the topography of such surfaces is not greatly different from abraded surfaces,

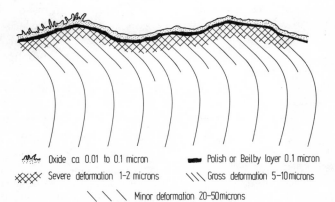

Oxide ca 0.01 to 0.1 micron   Polish or Beilby layer 0.1 micron

Severe deformation 1–2 microns   Gross deformation 5–10 microns

Minor deformation 20–50 microns

Fig. 1. Schematic diagram of a typical polished metal specimen showing topography and surface structure.

TABLE I

SOME TYPICAL SURFACE TOPOGRAPHICAL VALUES

Type of surface treatment	Surface parameters, micron (μm)		Surface topographical parameter, $(\sigma/\beta)^{1/2}$
	σ	β	
Bead blasted	1.4	13	0.33
Metallurgically polished	0.14	150	0.03
Run-in without seizure	0.06	240	0.016
Very finely polished	0.014	480	0.006

σ = mean deviation of asperity heights (for a Gaussian distribution, this is comparable with the center line average C.L.A.)

β = average asperity-tip radius

$(\sigma/\beta)^{1/2}$ = surface topographic parameter [see eq. (4)]. A low value is a measure of the smoothness of the surface.

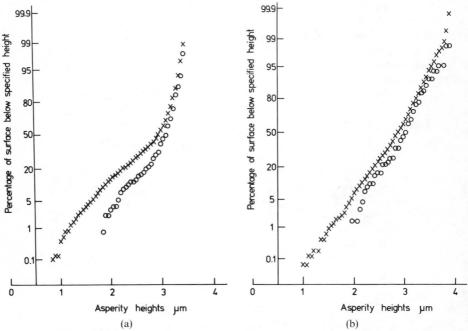

Fig. 2. Cumulative height distribution for a mild-steel surface prepared in two different ways. (a) Ground and lightly abraded. (b) Ground and lightly polished (Greenwood and Williamson, 1966). A perfectly Gaussian distribution would give a straight line. It is seen that, especially in case (b), both the distribution of all the height (x) and the peak heights (0) are very nearly Gaussian.

although the asperities may be on a finer scale and of a somewhat different geometry (Henzler, 1973).

Profilometer studies of real surfaces show that the topography can be conveniently approximated to a series of cones, pyramids, spherical caps, and grooves but that the average slopes of the asperities rarely exceed a few degrees. Of course at the atomic level projections of a few atomic steps may be regarded as making an angle of 90° to the surface. Mechanically prepared surfaces have a much coarser scale of roughnesses. With such surfaces the heights of the asperities depend on the type of surface treatment (see Table I), but the distribution of heights is roughly Gaussian, as may be seen from Figs. 2a and b. We now consider what determines the area of contact when such surfaces are placed in contact.

D. The Area of Contact Between Surfaces: Analytical Models

Instead of dealing with two rough surfaces in contact, we deal here with a rough deformable surface in contact with an ideally smooth hard surface, as shown in Fig. 3. The shaded area represents that part of the rough surface where true contact occurs. The analytical models we describe all assume ideal physical properties that extend to the outermost surface layers. They ignore surface films and the effect of size on strength properties. Both these factors may have a pronounced effect on the real situation as will be noted later.

1. PLASTIC DEFORMATION

The simplest model is for a rough metal surface. When placed in contact with a smooth hard surface, the tips of the asperities are deformed, at first elastically; for loads exceeding more than a minute value the elastic limit is exceeded and plastic flow occurs. It turns out that the local plastic yield pressure p_0 is very nearly constant and is comparable to the indentation hardness of the metal. Under these conditions the area of contact for any one asperity bearing a load w_1 is $A_1 = w_1/p_0$; thus for an assembly of

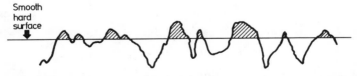

Fig. 3. Contact between a rough deformable surface and an ideally smooth hard surface. The load is supported by the shaded area, and the analysis ignores the problem of where the displaced material goes.

asperities the total area is

$$A = A_1 + A_2 + \ldots = \frac{w_1}{p_0} + \frac{w_2}{p_0} + , \ldots = \frac{W}{p_0}, \tag{1}$$

where W is the total load. The area is thus proportional to the load and independant of the size or smoothness of the surfaces. A more detailed study by Williamson and Hunt (1972) takes into account the interaction of the plastic zones between each asperity. This increases the effective plastic yield pressure; thus for extensive deformation the area of contact will be less than that given by Eq. (1).

2. Elastic Deformation

The earliest models dating from Parent (1704) assume that surfaces are covered with asperities that approximate to portions of a sphere. If these have a radius of curvature R, Young's modulus E, and Poisson's ratio v, the area of contact for an individual asperity under a load w_1 is

$$A_1 = \pi \left\{ \frac{3w_1 gR}{4} \left(\frac{1-v^2}{E_1} \right) \right\}^{2/3}. \tag{2}$$

For each asperity A_1 is proportional to $w_1{}^{2/3}$. If all the asperities, on average, have the same curvature, we may now consider two distinct ways in which elastic deformation can occur (Archard, 1957):

(i) The number of asperity contacts remains constant so that an increase in load increases the elastic deformation of each contact. The area of real contact is then proportional to $W^{2/3}$.

(ii) The average area of each deformed asperity remains constant and increasing the load increases the number of regions of contact proportionally. Clearly the real area of contact is directly proportional to W.

Archard concluded that in any real situation where elastic deformation occurs, the area of contact will be proportional to W^m, where m lies between $\frac{2}{3}$ and 1. This is supported by a more detailed analysis. For example, Lodge and Howell (1954) considered a spherical surface (radius R) covered with a close-packed array (n/cm^2) of hemispheres of radius r. For elastic deformation the true area of contact turns out to be

$$A = kr^{2/3} n^{1/3} R^{2/9} W^{8/9}. \tag{3}$$

If the asperities are covered with still smaller asperities, then, as Archard has shown, A becomes proportional to $W^{26/27}$. Thus a multiple asperity model involving purely elastic deformation tends to give an area almost linearly proportional to the load. Such a model would not give a power of W greater than unity. However, Mølgaard (1962) has shown that such a situation is possible over a limited load range if the contour of the surfaces has a suitable geometry. The contour is such that, in effect, there is a collapse from one asperity-covered level to a neighboring level as the load is increased.

Of course the two-thirds power in equation (2) follows from the assumption that the asperities are spherical. If they are conical or pyramidal in shape, the law of geometric similarity shows that, even for purely elastic deformation, the area of contact will be exactly proportional to the load. For example, for a cone of semiapical angle θ, the area of contact under a load w_1 is

$$A_1 = w_1 \frac{2(1 - v^2)}{E \cot \theta}.$$

Thus if all the asperities have the same value of θ, the total area A will be proportional to W. These conclusions would be applicable to rubber surfaces or to hard surfaces such as diamond and carbides at very small loads (see Section III).

3. ELASTIC AND PLASTIC DEFORMATION

A more realistic model for materials that can deform both elastically and plastically, e.g. metals, ionic solids, glasses, is that due to Greenwood and Williamson (1966). They first consider the behavior of surfaces for which the asperity heights follow an exponential distribution. Their analysis shows that whatever the law of deformation this will give an area of contact directly proportional to the load. Physically this means that the *average* size of the asperity contacts remains constant whatever the load, so doubling the load doubles the number of asperity contacts. Even if, say, $x\%$ of the contacts are plastic and the remainder elastic, the same conclusion follows. There will always be $x\%$ plastic. As the load increases, the existing contacts will increase in size, but new smaller ones will be formed so that the load–area proportionality will remain.

However, real surfaces do not show an exponential distribution. As was pointed out previously, the distribution is much more nearly Gaussian (Fig. 2). The deformation behavior now depends on two main parameters: (i) the surface topography, which can be described by the square root of the ratio σ/β, where σ is the mean deviation of the asperity heights and β the radius of the tips of the asperities (assumed spherical); and (ii) the deformation properties of the material as represented by the ratio of the elastic to the plastic properties. This can be written E'/p_0, where E' is the reduced Young's modulus[†] and p_0 is the pressure at which local plastic deformation occurs. Both parameters are dimensionless, and the product is termed the plasticity index

$$\psi = (E'/p_0)(\sigma/\beta)^{1/2}. \tag{4}$$

Greenwood and Williamson (1966) show that if $\psi > 0.6$, the deformation will be elastic over an enormous load-range. This corresponds to materials for which p_0 is large compared with E' and/or to smooth surfaces for which σ is small compared with β. In this regime the asperity contact pressure increases somewhat as the load increases, but the change is not large. The contact pressure is of order $0.3\,E'(\sigma/\beta)^{1/2}$ over a very wide range of loads, thus the true area of contact is again roughly proportional to the load. For very smooth surfaces the true contact pressure turns out to be between 0.1 and $0.3\,p_0$.

For most engineering surfaces $\psi > 1$; the deformation is now plastic over an enormous range of loads, and the true contact pressure is p_0. This is the simple model described above; the area of contact is again proportional to load as represented by eq. (1).

The elastic and plastic regimes are shown in Fig. 4 for a Gaussian distribution of roughnesses on a series of solids of different hardnesses

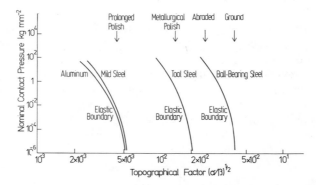

Fig. 4. Graph showing the nominal pressure at which the transition from elastic to the onset of plastic deformation occurs for a flat metal of specified roughness pressed on to a flat ideally smooth hard surface. The surface topography factor is represented by the ratio $(\sigma/\beta)^{1/2}$, where σ is the standard deviation of the asperity heights and β the radius of curvature of the asperity tips, which are assumed to be spherical. The asperity height distributions are assumed to be Gaussian. The materials shown are aluminum ($p_0 = 40\,\mathrm{kg\,mm^{-2}}$); mild steel ($p_0 = 120\,\mathrm{kg\,mm^{-2}}$); tool steel ($p_0 = 400\,\mathrm{kg\,mm^{-2}}$); ball-bearing steel ($p_0 = 900\,\mathrm{kg\,mm^{-2}}$).

[†] The reduced modulus is defined as $E' = E/(1-v^2)$ where E is Young's modulus and v Poisson's ratio for the deformable body.

(aluminum, $p_0 = 40 \, \text{kg mm}^{-2}$; mild steel, $p_0 = 120 \, \text{kg mm}^{-2}$; tool steel, $p_0 = 400 \, \text{kg mm}^{-2}$; ball-bearing steel, $p_0 = 900 \, \text{kg mm}^{-2}$). For a given surface finish (fixed σ/β) the deformation is elastic if the nominal pressure is below the curve. If the loading is increased so that it passes across the line, a fraction of the asperities will begin to deform plastically. This fraction will increase with increasing load until gradually the major part of the contact becomes plastic. It is seen that with aluminum a nominal pressure of $4 \, \text{kg mm}^{-2}$ (i.e., one-tenth that required to produce completely plastic deformation) will produce predominantly plastic deformation of the asperities for even the smoothest surface. With tool steel the same relative condition of contact (mean pressure $= 0.1 p_0$) occurs for surfaces rougher than metallurgical polish. Only with ball-bearing steel is the contact predominantly elastic even for relatively rough surfaces.

We conclude that for solids that deform elastically, the area of contact will increase as W^m, where m is between $\frac{2}{3}$ and 1; for multiple asperity surfaces the index will be nearer 1 than $\frac{2}{3}$. For ductile materials such as metals the deformation of the asperities for typical engineering surfaces will be mainly plastic, and the true area of contact will be given by

$$A = W/p_0. \tag{5}$$

For extremely smooth hard surfaces the deformation may be elastic, and the area of contact will be given by $A = \text{load}/p$, where p is of the order 0.1 to $0.3 \, p_0$. Recent simplified reviews of asperity models will be found in papers by Archard (1974) and Tabor (1975).

We may finally consider two other classes of materials: brittle solids and polymers. In simple tensile experiments brittle solids extend elastically and then pull apart at some critical tensile stress. If, however, the material is subjected to a high hydrostatic pressure, brittle failure is prevented and the material can deform plastically. It so happens that under conditions that exist between an asperity and another surface, the stress situation involves a large hydrostatic component. In may cases this is sufficient to inhibit brittle failure and the asperity deforms in a plastic manner, although some cracking may also occur. Consequently the behavior, to a first approximation, resembles that described above for ductile materials.

The behavior of polymers is more complex. They are neither elastic nor plastic but viscoelastic; thus the deformation is time-dependent as well as being dependent on load and geometry. However, over a fairly wide range the contact area between a hard sphere and a smooth flat specimen of polymer (or between a smooth sphere of polymer and hard flat surface) is given by $A = W^m$, where m is generally between 0.7 and 0.8. It is possible, by making a few reasonable assumptions (Cohen and Tabor, 1966) to incorporate this behavior into the Lodge and Howell model described previously. The area of contact A is then given by

$$A = k \frac{n^{1-m} r^{2-2m} W^{2m-m^2}}{G^{2m-m^2}} R^{2(1-m)^2}, \qquad (6)$$

where G is an effective modulus having the dimensions of pressure. It is seen that for $m = \frac{2}{3}$ (elastic case), Eq. (6) reduces to the Lodge and Howell equation and G is the elastic modulus. For $m = 1$ it reduces to an area of contact independent of R and proportional to W:

$$A = k(W/G). \qquad (7)$$

Equation (7) corresponds to plastic deformation, and G corresponds to the hardness or yield pressure of the solid.

E. Surface Films and Size Effects

There are two factors that greatly affect the contact between solid surfaces. The first is the presence of surface films—this is particularly important with metals whose surfaces are usually covered with oxide layers. These may undergo very heavy deformation at the regions of contact. If they are ductile, they will deform with the metal and retain their integrity, thus preventing intimate metal–metal contact. If they are brittle, they will crack and metal will tend to flow through the cracks themselves (Osias and Tripp, 1966). These films are especially important in adhesion and friction, but unfortunately it is not easy to specify or quantify their strength and yield properties.

The second factor concerns the effect of size on the strength properties of solids, in particular the strength properties of the individual asperities themselves. Size has little effect on elastic properties, but it can have a marked influence on brittle strength and plastic yielding. It is well known that brittle solids owe their weakness to the presence of flaws or cracks, and the smaller the specimen the less chance there is of including a large crack. Consequently the material becomes much more resistant to brittle failure, a phenomenon that is well known in fiber technology. A similar effect arises in relation to plastic yielding. If the volume being deformed is very small, it may not contain any mobile dislocations; and, in that case, as Gane (1970) has shown, the plastic yield stress may reach very high values representative of the "ideal" crystal lattice. Thus minute portions of a single crystal of gold may exert a resistance to indentation ten times larger than the material in bulk—p_0 acquires a value of order $200 \, \mathrm{kg \, mm^{-2}}$ compared with a value of about $20 \, \mathrm{kg \, mm^{-2}}$ for bulk gold. This implies that asperities may undergo much larger elastic deformation before plastic flow occurs, and the area of true contact may be appreciably less than that calculated from bulk values of p_0.

It should, however, be borne in mind that for work hardened metals or alloy-steels the bulk-hardness itself is relatively high, and the hardness of small volumes may not be more than a factor of, say, two greater than the bulk value.

There are no theoretical models of contact between solids that take these factors into account.

III. THE AREA OF TRUE CONTACT: EXPERIMENTAL

A. Introduction

If we place two polished flat surfaces, say 10 cm² in area, on top of one another, the geometric area of contact is 10 cm². However, the contact occurs only where the asperities on one surface touch the other. This is the basis of the analyses described in Section II, where the "true" area has been expressed solely in terms of bulk deformation properties of the solids. In this section we may ask if the true area of contact can be defined more precisely. The true area is that area over which the atoms in one surface are within the repulsive fields of those on the other; this is the way in which atoms "bear a load." There will of course be a penumbral region around each such contact region where the atoms are perhaps only twice as far apart. Since the repulsive forces are extremely short-range, these will not bear any of the load. On the other hand the attractive forces, although also short-range, do not fall off quite so quickly; thus in these penumbral regions there may be some attractive forces between the atoms. For typical engineering surfaces the attractive forces in these penumbral regions are small and play a negligible part in determining the area of true contact. But there may be circumstances (see below) where they are by no means negligible.

We now consider briefly the experimental methods available for determining the true area of contact.

B. Profilometer Methods

This is the exact analog of the profilometer methods used to describe surface topography. The modification is that the surface contour is first determined along some particular direction. Another surface is pressed on to it, removed, and the new profile determined. The flattened areas provide a fairly good indication of the regions where plastic contact has occurred. The method will not, however, indicate very directly regions where the original deformation was primarily elastic. Further, the resolution, although adequate for most engineering purposes, is very poor on an atomic scale.

C. Electrical Resistance Methods

With metals it is possible to deduce the contact area while the surfaces are actually in contact by measuring the electrical resistance (Holm, 1958). If two surfaces of the same specific resistivity ρ make perfect metallic contact over a circle of radius a, the electrical resistance R_1 is in the nature of a "constriction resistance" and has the value

$$R_1 = \rho/2a. \tag{8}$$

It is clear that if contact occurs over a number of asperities, the electrical resistance will not enable us to determine the area of contact unless this number is known. For example, if the contact area (πa^2) is divided into n equal circles each of radius $(a/n^{1/2})$ and the individual asperities are so far apart that the current flow through one contact does not interfere with that of the others, the total resistance will be the summation in parallel of n resistances each of value $\frac{1}{2}\rho/(a/n^{1/2})$. Thus although the total area of contact will still be πa^2 the contact resistance will be

$$R_n = R_1/n^{1/2}. \tag{9}$$

If the individual asperities are closer together, the current streams will interact and the resistance will be larger than this, the extent depending on the closeness of the asperities. For example, if the separation between the asperity contacts is comparable with the diameter of each individual asperity contact, the constriction resistance of the assembly is almost the same as that of a single junction embracing all the individual junctions. The area of true contact will thus be about four times smaller than that deduced from the resistance measurement.

There are two further complications. Due to the tunneling effect, electrons will readily cross gaps of the order of 10 Å so that an appreciable penumbral area will be included. Secondly, oxide films will greatly complicate the measurements. In spite of all these defects, electrical resistance measurements remain the simplest and most direct experimental methods of making some estimate of the area of contact between metal surfaces.

Some attempts have been made to use thermal conductance in a similar way. They are less satisfactory.

D. Ultrasonic Methods

Kendall and Tabor (1971) have described a method in which a pulse is injected into one of the bodies, and the acoustic energy transmitted through the interface into the other body is measured. This method can be used for all solids (not only metals), and it is not greatly affected by the presence of thin contaminant or oxide films, since the acoustic impedance of oxide films is not greatly different from that of metals. However, the acoustic transmittance of a small constriction is proportional to its diameter, not its area. From the point of view of quantitative determinations of area, it therefore suffers from the same defects as the electrical resistance method. More recent work employing a high frequency accoustic method has been described by Kragelsky et al. (1974).

E. Optical Methods

With nonmetals, if one or both of the surfaces is transparent, a useful experiment approach is the optical one. If the surfaces are molecularly smooth, the optical area determined by interference method will be identical with the true area of contact. The only material for which this is applicable is mica (Bailey and Courtney–Pratt, 1955). For other surfaces, which are not molecularly smooth, there is always some uncertainty as to the difference between the optical and the real areas of contact (Adams, 1963). Mica has an additional advantage: multiple beam interferometry can be combined with the technique of fringes of equal chromatic order. This gives a resolution of the order of 3 Å, so the penumbral area included in the optical area is negligible.

Other optical methods include phase contrast (Dyson and Hirst, 1954) and total internal reflection (Kragelsky and Sabelnikov, 1957). These methods involve some uncertainty, since the optical area includes a non-contacting penumbral region where light is able to cross a small gap of the order of a few hundred angstroms.

All these methods confirm the general view that when hard, nominally flat surfaces are placed in contact, the true area of contact is generally very much smaller than the apparent area. This is due to the presence of surface irregularities. The exception is the contact between two cleavage faces of mica if no cleavage steps are present. Another exception is the behavior of relatively smooth, soft elastic solids such as rubber, where the asperities are easily depressed into the hinterland (Roberts and Tabor, 1971).

IV. TYPES OF ADHESIVE FORCES

A. Metallic

If two *clean* pieces of gold are gradually brought together, the forces between them will follow the following pattern. For separations greater than, say, 20 Å they will be attracted by van der Waals forces, which increase as the separation decreases. At some separation the surface plasma will provide a screening action—at the same time the metallic bond will begin to develop. When the surfaces are at an atomic distance apart, the full metallic bond will have been formed and the short-range repulsive forces will also come into operation to provide final equilibrium between the two bodies.

A stimulating if somewhat controversial treatment of this type of inter-action in terms of bulk and surface plasmons has recently been given by Schmit and Lucas (1972a, b). They have also shown that these concepts can be successfully applied to transition metals if it be assumed that the d electrons are the primary cause of plasmon interaction. Another approach, rather qualitative in nature (Czichos, 1972), treats the metals as a "jellium" and suggests that the adhesive bond strength will depend on the free-electron density. For dissimilar metals adhesion is a function of the donor–acceptor tendencies of the two metals. Similarly, in a far more analytical attack Ferrante and Smith (1973) have attempted to describe adhesion in terms of electronic structure and overlap [see also Djafari Rouhani and Schuttler (1973)].

If the initial specimens expose the same crystal planes and are in the same orientation, the two pieces of gold should become a single piece over the area of atomic contact. The adhesive strength should then be equal to the bulk strength of gold itself. If the orientations are different, the interface will resemble a grain boundary in a single gold specimen. The strength of the interface will depend on the crystal planes and the orientation of these planes relative to one another. According to Czichos (1972), this strength depends on the extent to which the Fermi surface allows electron-bonding in a particular direction to occur. If there is marked mismatch, the application of a compressive or shear stress to produce plastic deformation at the inter-face will aid adhesion. Presumably it will increase the amount of atomic contact. In addition, as Semenhoff (1958, 1961) suggests, plastic deformation, like thermal energy, may activate the bonding process. On the other hand, Holm (1958) maintains that metallic bonds do not require activation.

For two dissimilar metals A and B, the same picture applies and the bond at the interface will generally lie between the bond strength of A and that of B. Thus if A is stronger than B, the bond will be weaker than A but stronger than B. If the surfaces are pulled apart, the bodies will tend to separate within B leaving fragments of B attached to A. If the work-functions of the metals are different, a contact-potential will be established across the interface, but there is little evidence to suggest that this adds appreciably to the strength of adhesion. On the other hand Buckley (1969) and Czichos (1972) suggest that strong adhesion will only occur if one metal can act as an electron donor and the other as an electron acceptor.

If an insulator is brought into contact with a metal, there will be a far larger separation of charge at the interface. This will produce an electrostatic attraction in addition to the van der Waals interaction between the bodies. Detailed experiments in which the insulator is a polymer have been described by Derjaguin (1955), who claims that practically the whole of the adhesion is electrostatic in origin. By contrast, Skinner et al. (1953) maintain that not more than 20% of the adhesion arises from this source.

Recently Phillips (private communication) has described in greater detail the interaction between a metal and an insulator or semiconductor. Charge-transfer, as described above, occurs, and if the dielectric constant of the non-conductor is below about 5, the charge separation produces a potential drop equal to the difference in effective work functions. If, however, the dielectric constant is greater than 7, a type of electrical double layer, probably associated with the presence of surface states (Bardeen, 1949, 1964), is generated at the interface canceling out the work-function potential. One might therefore expect stronger adhesion between a metal and a nonconductor of low dielectric constant (e.g., quartz or a polymer) than between a metal and a nonconductor of high dielectric constant (e.e., diamond, Si, Se, Ge). This aspect of adhesion, which really goes back to the very old problem of contact electrification, has not been examined experimentally in any great detail. However, in the last two or three years a few papers have been published showing that the adhesion to a semiconductor surface can be markedly changed by exposure to ultraviolet light. This exposure changes the effective Fermi level and the charge density of the electric double layer formed at the interface (Schnabel, 1969; Krupp, 1972, Derjaguin et al., 1972).

There is little doubt that the next few years will see a large number of theoretical papers attempting to interpret adhesion in terms of the band structure. Most of them will probably be qualitative; those that are quantitative are more likely to give the energy of bond formation and rupture rather than the forces involved. As in all strength measurements, there is often a wide gap between theoretical and observed values due to the presence of flaws, dislocations, etc. Thus bonds may be intrinsically strong, yet the *observed* adhesion may be relatively weak. On the other hand, this type of theoretical treatment may well explain one of the earlier paradoxes in this area: that although thoroughly clean metals (even if mutually insoluble) give strong adhesion, there are certain metal combinations (e.g., iron–vanadium) where the slightest amount of contamination drastically reduces the adhesion, because for these materials electron transfer across a very thin adsorbed film is greatly diminished.

B. Ionic

If two clean ionic crystals of the same material are brought together, the position is as follows. If the ionic charges are uniformly distributed over the surfaces (i.e., no clustering of individual charges), for separations greater than a few atomic dimensions a charge in one surface will see a smeared out positive-and-negative charge in the other; it will indeed appear electrically neutral, and the net Coulombic force will be very small compared with the van der Waals interaction. Only when the surfaces are close enough for the graininess of the charges to be apparent will the Coulombic forces become important. If the initial crystal planes and orientations are identical, say a (100) face of NaCl exactly matching a similar face on the other specimen, it could happen that the negative charges on one surface might be exactly over the negative charges on the other. The same would apply to the positive charges. In principle this would give a repulsive force. But the situation is unstable and energetically unfavorable. A minute shear strain would be sufficient to displace one surface relative to the other by an atomic (more correctly, ionic) spacing, and strong attraction would occur. When the separation equals the atomic spacing, the bond will resemble that within the bulk of the crystal. If the orientations are different, the situations will be far more complex. The adhesive strength will approximate that of a grain boundary.

C. van der Waals

Van de Waals forces are now well understood. With polar molecules they arise from dipole–dipole interactions. With nonpolar molecules they arise from the interaction of fluctuating dipoles in the individual atoms (London, or dispersion van der Waals forces). When the atoms are more than a few hundred angstroms apart, the correlation between the electron fluctuations is poor and the resulting force is known as the retarded van der Waals force: the force between two atoms is proportional to $(distance)^{-8}$. For separations less than 100 Å the correlation is close and the interaction is that of the classical London (1931) type: the force falls off as $(distance)^{-7}$. If two non-polar van der Waals solids are brought together, the van der Waals forces may be calculated by assuming additivity of the individual atoms (Hamaker, 1937) or by the macroscopic theory of Lifshitz (1956). The results are not greatly different and have been conveniently summarized in two recent review articles, one by Krupp (1967) and the other, more recently, by Israelachvilli and Tabor (1973). For two parallel surfaces, d apart, the force per unit area is

$$f = A/6\pi d^3 \qquad (10)$$

for separations smaller than 100 Å, where A is the Hamaker constant. The value of A depends on the polarizability of the solid but does not vary by a factor of more than 10 for the most diverse materials. For purely van der Waals solids it is of order 10^{-12} ergs(10^{-19} J), for metals of order 10^{-11} ergs (10^{-18} J).

Although this book is concerned primarily with crystalline solids, the behavior of crystals of argon or krypton is of little practical interest. A far more important type of van der Waals material is represented by rubbers or polymers. Here the interaction energy can be calculated satisfactorily using Eq. (10) right down to atomic contact, even though at the final stage the atomic graininess of the material is apparent (Mahanty and Ninham, 1973). There are four other factors involved with polymeric materials. First, inter-diffusion of polymeric chains across the interface may occur. This will greatly increase the adhesive strength, since valence bonds, as distinct from van der Waals bonds, will be established (Voyutski, 1963). Secondly, for dissimilar materials charge-separation may lead to an appreciable electrostatic component (Davies, 1973). Thirdly, these materials are easily deformed by comparison with the other solids discussed in this section. With soft rubbers for example, large areas of intimate contact can easily be established; consequently, although the interfacial forces themselves are weak, it is not difficult to obtain relatively high adhesive strengths. A similar factor probably accounts for the strong adhesion between sheets of thin polymeric films. Finally, adsorbed films appear to have less effect in reducing the adhesion of polymers than is the case with metals.

D. Covalent

The forces between covalent solids are more difficult to specify. When such solids are brought into intimate contact, one might expect the bonding across the interface to be similar to the bonding within the solid. However, there is some evidence that the bonds on the free surface are relaxed and that a finite amount of energy is required to activate them (Holm, 1958).

There is an additional factor that is the inverse of that discussed in relation to rubber-like materials. Most covalent solids have a high elastic modulus and, if they can be deformed plastically, are generally extremely hard. Consequently, in practical experiments it is often difficult to obtain large areas of contact (because molecularly smooth surfaces are usually unobtainable) even if appreciable joining loads are employed. Further as the joining load is removed elastic recovery (see V.E.1) may break the adhesional junctions. Consequently, the adhesion force *measured* may be much smaller than the true adhesion occurring while the joining load was originally applied.

V. EXPERIMENTAL STUDY OF ADHESION

The experimental study of adhesion usually involves three separate factors: the cleanliness or otherwise of the surfaces; their surface topography; and their deformation properties. In addition the observed adhesion may depend markedly on the method of measurement.

A. Methods of Measuring Adhesion

In the adhesives industry a great deal of effort has been mounted in devising nondestructive adhesive tests. These attempts have not been particularly successful, although some have proved useful technologically. Usually they involve an acoustic measurement, and, in effect, they establish the existence or otherwise of solid continuity across the "glue-line."

Basic adhesion measurements imply the application of a force sufficient to separate the adhering surfaces. In this sense they must be destructive. The surfaces are attached to grips and inserted in a suitable tensometer. The two major factors are, first, that the force must be normal to the interface, and second, that the force must be applied at a specified uniform rate. In some tests the adhesion may be measured by applying a force tangential to the interface. In both types of experiment uncertainties arise due to stress concentrations in the contact area and to the nonalignment of the separating force. These two factors will tend to peel the surfaces apart rather than to pull or shear them apart.

If one of the adhering surfaces in in the form of a thin film (rather than a solid body), the adhesion may be measured by deliberately attempting to peel the surfaces apart. The film is attached to a suitable grip and pulled away in a direction normal to the interface. The method is widely used in the paint and adhesives industry, but it is well recognized that the results are a function of many variables apart from the "true" adhesion itself. It is of greatest value as a comparative rather than an absolute test. Another approach is the scratch test, in which a stylus is slid over the film and the normal load increased until the film begins to separate from the substrate (see Section V.F).

Other methods have been used for applying a normal force across the interface. For example, the adhering bodies may be shot at a high velocity in a direction normal to the interface and one of them suddenly arrested. A more elegant method is the use of a centrifuge. For example, Krupp (1967) has studied the adhesion of gold particles to a steel rotor that is spun at an increasing speed until the particles fly off. In the detergent industry the adhesion of fine particles has been measured by increasing the rate of flow of a liquid over the particles until the hydrodynamic forces are sufficient to remove them.

It is evident that both at a fundamental and at an applied level the measurement of adhesion leaves much to be desired.

B. Metals

1. SIMILAR METALS: DUCTILITY AND RELEASED ELASTIC STRESSES

The adhesion between fairly clean metals was first studied critically by Bowden and Rowe (1956), although in their work the surface treatment and the vacuum used (10^{-6} Torr) were far inferior to those employed by later workers. In most of their experiments they pressed a hemisphere on to a flat with a known force and then measured the force required to pull the two apart. Broadly speaking, they found that most clean metals would stick strongly to one another though the adhesion always seemed to be less than that expected from simple considerations discussed previously. For example, if two clean pieces of silver are pressed together with a load W to produce plastic deformation at the contact region, the area of true contact will be of order W/p_0, where p_0 is the yield pressure of silver. On general grounds one would expect the interface to have a strength comparable with that of bulk silver so that the pulling process should be approximately the reverse of the joining process. The adhesional strength Z should therefore be comparable with W. It was always appreciably less, and Bowden and Rowe

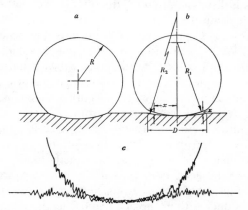

Fig. 5. Model showing the role of released elastic stresses in the adhesion of metals. (a) A smooth sphere is pressed on to the surface of a smooth flat surface to give plastic contact over a circle of diameter D. (b) If there is no adhesion, when the joining load is removed both surfaces will recover elastically and produce a change in geometry at the interface. (c) Contact between real surfaces showing asperities and asperity junctions.

attributed this to the effect of released elastic stresses. This is illustrated in Fig. 5. Figure 5a shows an ideally smooth sphere making plastic contact over a region of diameter D. When the joining load is removed, elastic stresses in the hinterland are released and tend to produce a change in geometry at the interface as shown in Fig. 5b; this in turn produces infinite stresses at the periphery of the contact region. The peripheral junctions will therefore be torn apart unless they are sufficiently ductile to accommodate themselves to the new geometry of the separating surfaces. This may be achieved at room temperature with metals such as lead or indium, which are self-annealing and extremely ductile; they give very high adhesive strengths. On the other hand, with metals such as silver the work hardening that is produced around the contact zone by the loading process itself tends to limit the ductility, and the adhesion is low. If the temperature is raised, the junction material can be annealed and strong adhesion is obtained. This explains why the sintering of metals is more effective if carried out at higher temperatures. Furthermore, high temperatures accelerate interfacial diffusion; in addition they favor surface diffusion that leads to mass transport of metal to the interface with a consequent increase in the area of intimate metallic contact (Easterling and Swann, 1971, Navara and Easterling, 1971).

Surface roughness is also an important factor, and if the roughnesses are large compared with the geometric contact area (see Fig. 5c), this will increase the tendency of the surfaces to pull themselves apart when the joining load is removed.

It is clear from this discussion that the ductility of the metals is important, particularly if the loading is sufficient to produce plastic deformation. This is again emphasized in a somewhat different way in the experiments of Dayson and Lowe (1972). Even under conditions of elastic loading, ductility may determine the extent to which the junctions will withstand a tensile stress before failing. For example, Pfaelzer (1971) carried out some very delicate adhesion experiments in a vacuum of 10^{-10} Torr between crossed cylinders of clean metals. The joining load ranged from about 2 g down to a few milligrams so that under these conditions the deformation was elastic. Since contact occurred over a fairly well-defined region, he was able to estimate the area of contact using electrical resistance methods. He found that the adhesion of thoroughly clean cobalt to itself, although appreciable, was weaker than the adhesion of copper to itself. As a hexagonal close-packed metal, cobalt is less ductile than the bcc metal copper. He also found that mutual orientation had little effect on the adhesion. Similar adhesion experiments between clean *flat* surfaces have been described by Buckley (see next section), though he reported some variation of adhesion with mutual orientation.

2. DISSIMILAR METALS: MUTUAL SOLUBILITY

At one time it was considered that mutually insoluble metals would show poor adhesion (Keller, 1963). However, it would now seem that if the surfaces are thoroughly clean, mutual solubility is not an important factor in influencing adhesion (Keller, 1967, 1972). On the other hand, it is clear that small amounts of contaminants may be much more effective in reducing the adhesion of some metals than of others. For example, a very small amount of oxygen (perhaps enough to give a monolayer) can produce a marked reduction in the adhesion of iron, whereas far more oxygen is required to produce a comparable reduction in the adhesion of copper.

A very thorough and elegant series of experiments on the adhesion of clean metals is due to Buckley (1971). He has been able to prepare clean flat surfaces of metals in a high vacuum apparatus, characterize the surfaces with LEED and Auger spectroscopy, place them in contact under a small load (~ 20 dyn), measure the adhesion, and then characterize the surfaces after they have been pulled apart. Indeed the only defect of this investigation is that, because of the geometry employed, he has little idea of the area of contact; it is probably extremely small. His results show that the adhesion does not depend on mutual solubility. In his view the parameter that most directly correlates with the observed adhesive is the cohesive energy. Some typical results taken from his paper, which show this correlation for the adhesion of various metals to clean iron, are given an Table II. In this table we have also included the surface energies γ of the metals, and this too correlates about as well with the adhesion. This is not surprising, since both the cohesive and surface energy are measures of the strength of the inter-atomic forces. However, there are two unsatisfactory features in this attempted correlation. First, if the adhesion is to be correlated with energies, the most appropriate quantity should be

$$\gamma_{iron} + \gamma_{metal} - \gamma_{iron-metal}.$$

Unfortunately, there is no direct data on the interfacial energy $\gamma_{iron-metal}$. Second, it is seen that the ductile metals lead and aluminum show a stronger adhesion than platinum. It would seem that in this part of his work, Buckley has underestimated the importance of ductility.

In general, but not always, he finds transfer of the softer metal to the harder. Buckley (1972) has also described some unique experiments in which oxidized tantalum and oxidized nickel were placed in contact with the (100) surface of clean iron. After the adhesion experiments AES showed that *oxygen* had been transferred to the iron surface.

Finally Buckley's experiments showed that with some alloys preferential segregation of one of the constituents could occur at the free surface.

TABLE II

ADHESION OF THE DENSEST PACKED PLANES OF VARIOUS METALS TO THE (011) SURFACE OF IRON

Metal	Cohesive energy J/g atom	Surface energy mJ m^{-2}	Solubility in iron at %	Adhesional force dyn (10^{-5}N)
Iron	40.5×10^4	1800	100	>400
Cobalt	42.6	1800	35	120
Nickel	42.9	1700	9.5	160
Copper	33.8	1300	<0.2	130
Silver	28.6	900	0.13	60
Gold	36.6	1200	<1.5	50
Platinum	56.4	1800	20	100
Aluminum	32.3	1000	22	250
Lead	19.7	500	—[a]	140
Tantalum	78.1	2200	0.20	230

Applied load 20 dyn (20 × 10^{-5}N); temperature 20° C; Ambient pressure 10^{-10} Torr.
[a] Insignificant.

Consequently, the adhesional behavior could be determined by the presence of relatively small quantities of what may be a minor constituent or even an impurity.

3. MUTUAL ORIENTATION

Semonoff (1958, 1961) has carried out a sustained study on the effect of mutual orientation in the adhesion of clean metals. His experiments show that if there is marked mismatch, the adhesion will be small. The adhesion can be increased by providing energy at the interface, either thermally (as mentioned previously) or by the work of plastic deformation.

4. METALS AND COVALENT SOLIDS

Experiments by Gane (1973) show that the adhesion between clean covalent solids may be very small but that the adhesion of metals to these solids can, in some circumstances, be very large. This is discussed in greater detail in Section V.E.2.

5. METALS IN AIR

The adhesion between metals in air has been studied by Anderson et al. (1957), Anderson (1960), and Sikorski (1964). If heavy deformation is imposed on the surfaces to break up surface films, strong adhesions can be obtained with many combinations. Some pairs, such as iron–vanadium, gave poor adhesion (see Section IV.A). Again, the self adhesion of a wide range of metals seems to fall into fairly well-defined groups depending on structure; for example, hexagonal metals form a self-consistent, poorly adhering group. The general conclusion seems to be that all surfaces will show strong adhesion but that some combinations of materials are more sensitive to surface contamination or to the effect of released elastic stresses. The importance of released stresses, emphasized by Rowe, is consistent with the observation that many practical adhesives fail, not because the adhesion is poor, but because stress concentrations in the "glue-line" provide a region of weakness (de Bruyne, 1956).

C. Ionic Solids

There is little work of a systematic nature on the adhesion of ionic solids. Some simple experiments 20 years ago by King and Tabor (1956) showed that strong adhesion could be achieved between freshly cleaved surfaces of rock-salt in air. Thin square slabs 5×5 mm across exposing the (100) face and 1 mm thick were placed in contact with their $\langle 100 \rangle$ directions parallel. They were inserted between hard plattens and subjected to a compression sufficient to double their area. Thus large scale plastic flow occurred and because of the constraining effect of the anvils, which ensured a relatively large hydrostatic stress, very little cracking took place. When removed from the anvils, the slabs were found to adhere very strongly together, the interface withstanding a tensile stress comparable to the tensile strength of a single crystal of rock-salt. If the crystallographic directions of the two slabs were out of register, the adheson was weaker even after appreciable plastic deformation. The weakest adhesion occurred when the $\langle 100 \rangle$ directions in the slabs were at 45° to one another. This may be due to unfavorable matching of the ionic charges or to differential elastic contractions in the interface when the compressive load is removed. However, the most likely explanation is that intersecting slip lines at the interface break up the contact area.

Strong adhesion was also observed between a hemisphere of indium and a clean cleavage rock-salt (100) face in air. When the indium was pulled off, a visible patch of indium was left attached to the rock-salt surface. Other work on the interaction between ionic solids is described in Bowden and Tabor (1964, Chapters VII and VIII).

D. van der Waals Solids

1. INORGANIC CRYSTALLINE SOLIDS

Although studies have been described on the friction of solid krypton (Bowden and Rowe, 1955), no systematic adhesion experiments have been carried out on this material. Studies have been made on the adhesion of solids such as ice (Raraty and Tabor, 1958), where the main interaction corresponds to van der Waals dipole–dipole forces and hydrogen bonding. Similarly, there have been many observations of the adhesion between cleaved mica surfaces. For the freshly prepared specimens there are strong surface charges due to the presence of potassium ions in the cleavage planes so that the interaction is Coulombic. The surface energy of such cleavage planes is of order 5000 mJ m^{-2} (5000 ergs/cm^2). When exposed to the air, these charges leak away or are neutralized by charges picked up from the atmosphere. The surface energy is of order 220 mJ m^{-2} if the surfaces are well matched in orientation and about 120 mJ m^{-2} if the orientations do not match (Bailey and Kay, 1967). The force between two such mica surfaces, if they are more than 10 Å apart, is mainly due to the London or dispersion type van der Waals interaction. When brought into atomic contact, however, a considerable short-range force is exerted; this may well be due to the presence of an adsorbed monolayer of water on each surface. On the other hand, Metsik (1973) believes that the potassium–oxygen ions form strong dipoles within the mica and that the main source of surface energy is the attraction of these "package" dipoles. He calculates a value of 300 mJ m^{-2} for the contribution to the surface energy from this source. The role of electrostatic forces has also been emphasized in a recent paper by Bailey and Daniels (1972). These factors and the effect of mutual orientation deserve further study.

2. ORGANIC VAN DER WAALS SOLIDS

Direct measurements of surface forces have recently been carried out between mica surfaces, each of which was covered with a condensed film of a long chain fatty acid (Israelachvili and Tabor, 1973). The forces have been determined for separations ranging from 14 to 240 Å. For separations greater than about 100 Å the behavior is dominated by the properties of the mica substrate. For separations less than obout 50 Å the behavior is dominated by the behavior of the hydrocarbon films themselves. From this it has proved possible to deduce directly the Hamaker constant of the hydrocarbon for short-range van der Waals forces; the value is about 10^{-12} erg or 10^{-19} J. This leads to two interesting conclusions. If the forces obey the same power law down to atomic separations, the energy gained in bringing two unit surfaces together (in principle, from infinity) is simply twice the surface energy γ. Hence

$$2\gamma = \int_{x_0}^{\infty} \frac{A}{6\pi x^3}\, dx = \frac{A}{12\pi x_0^2}. \tag{11}$$

Putting $A = 10^{-19}$ J, $x_0 = 2$ Å, we obtain a value for γ of about 30 mJ m^{-2}. This is a reasonable value for a hydrocarbon and agrees well with values deduced from critical wetting tensions (Zisman, 1963). The second conclusion is that at a separation of 2 Å the attractive force per unit area for parallel flat surfaces is $10^9\,\mathrm{N\,m^{-2}}$ or $10{,}000\,\mathrm{kg\,cm^{-2}}$. This would therefore be the force required to pull the layers apart. Even allowing for the short-range repulsive forces, this value would only be reduced by a factor of 2 or so. Thus the van der Waals forces could provide a tensile strength of a hydrocarbon of order of a few thousand $\mathrm{kg\,cm^{-2}}$. The observed strength of such a material is far lower, and the same discrepancy applies to the adhesive strength of a bond formed with, say, a hydrocarbon wax. The reason is that the real strength is determined by the presence of flaws, cracks, or local impurities that act as stress concentrators.

3. HIGHLY ELASTIC VAN DER WAALS SOLIDS

Although this treatise is concerned with crystalline solids, the adhesion of rubber-like materials sheds considerable light on certain aspects of the adhesive process. If a very smooth rubber sphere is brought very close to another surface, van der Waals forces will draw the surfaces together, and a finite contact area will be established even if no normal load is applied. A finite force will then be required to pull the surfaces apart.

A detailed calculation in terms of surface forces is difficult. A simpler approach is to use concepts of energy. Consider, for example, the approach of a rubber sphere to a hard, flat surface. Figure 6a shows the surfaces just making contact if there were no surface forces, Fig. 6b shows the shape of the contact as a result of these forces. Contact has been formed over a circle

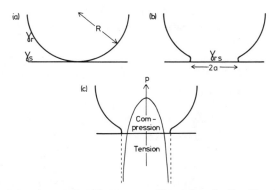

Fig. 6. Contact under zero load between a rubber sphere (radius R) and a hard, flat surface. (a) Contact if there were no surface forces. (b) Contact as a result of surface forces; the surfaces are drawn together over a circle of radius a. (c) The pressure distribution within the circle of contact; the central region is under compression, the peripheral regions under tension. (From Johnson et al., 1971.)

of radius a. The surface energy of the rubber has been diminished by $\pi a^2 \gamma_r$, that of the solid by $\pi a^2 \gamma_s$; while the interfacial energy has been increased by $\pi a^2 \gamma_{rs}$, where γ_r, γ_s, and γ_{rs} represent the surface energies of rubber, solid, and interface, respectively. The contact grows to the equilibrium diameter $2a$ when the surface energy surrendered, $\pi a^2 (\gamma_r + \gamma_s - \gamma_{rs})$, is exactly equal to the stored elastic energy in the rubber. The detailed calculation, which is the reverse of the classical Griffith fracture criterion, has been given by Johnson et al. (1971). As one might expect, the value of a is a function of the elastic modulus, of the radius of the sphere, and of the surface energies; and the results are in good agreement with theory. If a tensile force is now applied, the surfaces will pull apart at the peripheral regions; this will continue as the tensile force is increased until a stage is reached at which the rate of release of elastic strain energy just exceeds the rate of creation of new surface energy. This will mark the maximum tensile force the surfaces can withstand, and they will completely pull apart. It turns out that for spherical surfaces this maximum adhesive force Z is *independent* of the elastic modulus and has the value

$$Z = \tfrac{3}{2} \pi R (\gamma_r + \gamma_s - \gamma_{rs}), \tag{12}$$

where R represents the effective radius of curvature. [For example, for two spherical surfaces of radii R_1 and R_2 in contact, the value of $R = R_1 R_2/(R_1 + R_2)$.] For two bodies of identical rubber we may put $\gamma_r = \gamma_s = \gamma$, while for the interface we may assume as a close approximation that $\gamma_{rs} = 0$. Hence $Z = 3\pi R \gamma$. Some results for a rubber sphere ($R_1 = 2.2$ cm) in contact with a rubber flat are given in Fig. 7. The sudden fall-off in a at a critical value of Z is clearly shown. This critical value of Z is ~ 750 dyn; thus from Eq. (12) we obtain a value of $\gamma \simeq 34$ mJm^{-2}. This is a very reasonable value for the surface energy of a van der Waals solid.

The above analysis includes two simplifying assumptions. First, that the surfaces are so smooth that they make molecular contact over the whole of the region shown in Fig. 6. If the surfaces are initially of optical quality and if the modulus of the rubber is very low, small protrusions are easily squeezed down to a common level and this assumption becomes reasonably valid. This is probably one of the reasons why very soft rubbers generally appear to be tacky. If the surfaces are rough and/or hard, true molecular contact will occur over a smaller area within the macroscopic region of Fig. 6. Thus the amount of surface energy released in the contact process will be less than $\pi a^2 (\gamma_r + \gamma_s - \gamma_{rs})$, although the elastic energy stored in the hinterland will still be determined by the macroscopic value of a. This must necessarily mean that the surface will be deformed less, i.e., a will be less than for perfectly smooth surfaces. The pull-off force Z will also be less. A systematic study of the effect of hardness and roughness on adhesion has

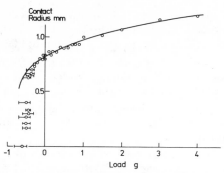

Fig. 7. Radius a of the contact zone formed between a rubber sphere ($R_1 = 2.2$ cm) and a rubber flat as the initial joining load of 4 g is gradually reduced and then made negative. The contact radius a remains finite until at a critical tensile force Z of about 750 dyn, it suddenly falls to zero as the surfaces pull apart.

only recently been completed (Gane *et al.*, 1974; Fuller and Tabor, 1975) and is discussed in Section E below.

Secondly, it is assumed that the deforming solids are ideally elastic. In fact some energy will be lost by hysteresis in the rubber, thus less of the released elastic energy will be available for separating the surfaces. Recently, in an entirely different type of adhesion experiment Andrews and Kinloch (1973) have studied the adhesion of rubber-like materials and have shown that the higher the hysteresis loss the greater the measured adhesive strength.

A third factor that the Andrews paper discusses is the possibility of chemical bonds at the interface. Yet a further factor is that emphasized by Voyutski. With polymeric materials molecular diffusion across the interface can lead to the formation of very strong bonds, corresponding to the strength of the carbon backbone.

E. Covalent and Other Hard Solids

This section describes work that has not yet been published and therefore contains somewhat more detail than is customary in a review.

The simplest covalent crystalline solid is diamond. In studying the adhesion of diamond there are two serious experimental difficulties. First, that of preparing atomically clean surfaces, since simple heating in a high vacuum may graphitize the surface if the heating is too severe. Surface treatments such as ion bombardment or gentle heating in hydrogen offer alternative methods, but these too introduce uncertainties. Second, it is not possible to obtain atomically smooth surfaces. The best mechanically polished surfaces are necessarily rough on an atomic scale. Experiments carried out on crossed cylinders of diamond ($R \simeq 5$ mm), pressed together with a force of a few grams weight in a vacuum of 10^{-10} Torr show zero adhesion whether the surfaces are cleaned by heating in vacuum or in hydrogen to $900°$ C or by argon ion bombardment (Gane *et al.*, 1974). According to the theory described in the previous section for elastic solids, the adhesional force, of order $\pi\gamma R$, should have a value of several grams weight, since γ for diamond is of order 5000 mJ m^{-2}. Since the apparatus could detect adhesional forces as small as a few milligrams (see Pfaelzer, 1971), the observed results are at least 1000 times smaller than expected.

Similar results have been observed for crossed cylinders of both chemically and mechanically polished sapphire. With crossed cylinders of TiC (which, although covalent, is a reasonable conductor of electricity) some adhesion was observed at room temperature, but it was very small, about 10 mg, whether the joining load was 10 mg or 2 g. Similarly, glass surfaces showed a small adhesion that was independent of the applied load. For these materials the contact at the loads employed must presumably have been elastic, thus the discrepancy with theory is very large.

Experiments show that with these hard nonductile materials the adhesion can be increased if very careful precautions are taken to reduce the effect of mechanical vibration. However, even when mechanical vibration is made as small as possible, the adhesion remains an order or even two orders of magnitude less than that given by the elastic analysis described above. There are three possible factors responsible for this: the effect of surface roughness, the need for thermal activation of the interfacial bond, and the lack of ductility of the junctions.

1. SURFACE ROUGHNESS

In order to study the effect of surface roughness, experiments were carried out on the adhesion between an optically smooth rubber surface and a hard flat optically smooth surface that could be roughened to various degrees. These experiments show that the adhesion observed on the smooth surface agrees well with theory but that it falls to a very low value for a surface roughness as small as a center-line-average (cla) of 1 μm. Dr. K. L. Johnson, in analyzing the somewhat simpler problem of flat surfaces in contact, points out (private communication) that the crucial factor is the *distribution* of asperity heights. For if the surface roughness consists of perfectly regular projections of equal height, the only effect would be to reduce the value of R in Eq. (12) for each contact, and the adhesion would correspond to the sum of the adhesions occurring at each of the individual asperities. However, if there are a few high asperities, they will prevent contact with the lower asperities and thus drastically reduce the total adhesion. If r is the average radius of curvature of individual asperities, z the deviation from the mean height, and E the reduced modulus for the two surfaces, the critical parameter is the dimensionless quantity

$$\theta = \frac{(\gamma_1 + \gamma_2 - \gamma_{12})\,r^{1/2}}{Ez^{3/2}}, \tag{13a}$$

where γ_1 is the surface energy of one surface, γ_2 of the other, and γ_{12} of the interface. This may be written more meaningfully in the form

$$\theta = \frac{(\gamma_1 + \gamma_2 - \gamma_{12})\,r}{Ez^{3/2}r^{1/2}}. \tag{13b}$$

It is obvious from Eq. (12) that the numerator is a measure of the adhesive force. Similarly, an examination of the Hertzian equations shows that the denominator is a measure of the force exerted by the asperities in pushing the surfaces apart. If θ is large, this implies that the adhesion is dominant; if θ is small, the asperities dominate and the adhesion is small. From the experiments with rubber ($\gamma_1 + \gamma_2 - \gamma_{12} = \Delta\gamma \simeq 30\,\mathrm{mJm}^{-2}$) the transition from strong to weak adhesion occurs for a value of $\theta \simeq 0.1$. Talysurf traces of the hard surfaces used in the present experiments give values of $r \simeq 0.15\,\mathrm{mm}$, $z \simeq 10\,\mathrm{nm}$. Assuming $E \simeq 10^{11}\,\mathrm{nm}^{-2}$ and $\Delta\gamma \simeq 1000\,\mathrm{mJ\,m}^{-2}$, we obtain $\theta \simeq 10^{-1}$. This would just enter the region of weak adhesion.[†] With harder materials such as diamond ($E \simeq 10^{12}\,\mathrm{nm}^{-2}$) surface irregularities of only a few atomic dimensions ($z \simeq 1\,\mathrm{nm}$) would be sufficient to reduce adhesion to a very low value. Evidently, with hard elastic solids surface irregularities are of major importance. Clearly, if the junctions can extend plastically when the load is removed, there is a much greater likelihood of adhesion dominating over roughness.

2. THERMAL ACTIVATION AND DUCTILITY

In order to see if the adhesion could be increased by "thermally activating" the bonds, the sapphire surfaces were pressed together at a temperature of up to 900° C, but no adhesion was observed. This temperature was the upper limit of the apparatus and corresponds to about $\frac{1}{2}T_m$ (where T_m is the absolute melting point) for sapphire, about 0.36 for TiC, and less for diamond.

By using crossed cylinders of Germanium ($T_m \simeq 1230°\,\mathrm{K}$) it proved possible to carry out adhesion experiments up to temperatures approaching the melting point. For a joining load of $\frac{1}{2}\,\mathrm{g}$, which should ensure a contact

---

[†] In principle a finite adhesion should always be observed corresponding, in the extreme case, to adhesion at a single projecting asperity. In the experiments with TiC this should give a minimum adhesion of $\frac{3}{2}\pi r\Delta\gamma$ or 0.5 nm. The much lower observed value, 0.05 nm, probably implies the presence of even finer-scale asperities.

pressure well within the elastic range, no adhesion was observed at room temperature. If the temperature was raised and the adhesion measured at the elevated temperature, the adhesion began to rise rapidly above 400° C ($\sim 700°$ K), corresponding to a T/T_m ratio above about 0.6. At a temperature of 700° C ($\sim 1000°$ K) the adhesion was about equal to the joining load. If, however, the surfaces while remaining in contact were allowed to cool to room temperature, the adhesion was small and irreproducible (see Fig. 8). Evidently the small adhesion is associated with limited ductility. Germanium is extremely brittle at room temperature and begins to show ductility only above obout 400° C. At higher temperatures it softens appreciably. Nevertheless, at the small joining loads used in these experiments the deformation in the contact zone during loading is primarily elastic over most of the temperature range examined. The important factor is the ductility of the material as the junctions are being pulled apart.

The importance of ductility is also shown by the following experiments. Although TiC adheres only weakly to TiC, copper adheres as strongly to TiC as it does to another specimen of copper. Similar results are observed with iron. Hydrogen has no effect on the adhesion of copper or iron on TiC, a result in agreement with earler work of Keller (1967) and Buckley (1966).

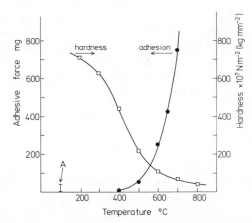

Fig. 8. Adhesion of clean germanium surfaces (crossed cylinders) in a vacuum of 10^{-10} Torr. The circles refer to experiments in which the surfaces were pressed together with a joining load of 0.5g at various temperatures and the adhesive strength measured at the same temperature. The adhesion is very small below 400°C, the temperature at which germanium becomes brittle. The circles refer to a single series of experiments in which the surfaces were pressed together at 700°C and the adhesion measured both at 700°C and at 200°C. It is seen that although the junctions formed give an adhesive strength of 0.75g when the adhesion is measured hot, the same junctions give an adhesive strength of only about 0.03g when the adhesion is measured at 200°C, as shown by point A. The variation of hardness of germanium with temperature ($\square$) is included for comparison, the data being taken from Trefilov and Mil'man (1964).

On the other hand, the admission of a small amount of oxygen produces a large reduction in the adhesion of iron and a much smaller effect with copper. It is probable that with copper, under the conditions of the experiment, no oxide is formed. However, with iron, Melmed and Carroll (1973) showed that small nuclei of oxide occur under conditions comparable with those obtaining in the adhesion experiments. These limit the amount of metallic contact and also provide bridges of limited ductility.

We may also quote the results of Anderson et al. (1957), who studied the adhesion between gold and germanium. In their experiments the gold was deformed plastically and the adhesion was greatly increased if the surfaces were heated to 250° C to restore ductility to the work-hardened gold.

Although this section has emphasized the importance of ductility, it is not the whole story. The adhesion of copper to clean sapphire or diamond is an order of magnitude smaller than that of copper to TiC. Presumably the interfacial bond between a metal and these largely covalent solids is fundamentally weak. By contrast the adhesion to TiC, which in terms of electrical conductivity is semimetallic (conductivity $\sim 100\,\mu\Omega\,cm$), is as strong as the adhesion of copper to itself.

F. The Adhesion of Thin Films

There is one aspect of adhesion where problems concerning the area of true contact do not arise, namely the adhesion of thin films deposited by evaporation or electroplating. In solid state technology evaporated films play a very important part, and a vast amount of research has been carried out on the nature of these films, their structure, the relation of their structure to that of the substrate (epitaxy), their electrical properties, etc.

Because of the experimental difficulties, little work has been carried out on the adhesive strength of such films. The most direct approach is to stick, solder, or cement to the film a suitable block, which can be attached to a tensometer (Belser and Hicklin, 1956; Butler, 1970; Lin, 1971). Beams (1956) attempted to use an ultracentrifuge to detach an electroplated film from the steel rotor. The rotor usually disintegrated before detachment occurred unless the interface was deliberately weakened with contaminants. An interesting technique developed by Heavens (1950) and by Weaver (Benjamin and Weaver, 1960, 1961, 1963) uses a scratch test in which a smoothly rounded point or stylus is drawn across the surface of the film. The normal load on the stylus is gradually increased until a load is reached at which the film under the stylus is detached and a clean scratch produced. Although the underlying mechanism is by no means clear (see Benjamin and Weaver, 1960; Kendall, 1971; Chen and Flavin, 1972), the method is extremely convenient as a comparative test and has even been used as a quantitative tool. For example, Benjamin and Weaver deduced that the adhesion of evaporated metal films on to the surface of the alkali halides could be explained quantitatively in terms of van der Waals forces. Using the scratch technique, Weaver (1972) found that the adhesion of aluminum to glass was much stronger than that of gold to glass. He then studied the adhesion of a gold film overlaid with aluminum and showed by the scratch method that the adhesion increased with time as aluminum diffused to the glass–metal interface. A major difficulty in interpreting such results is that vacuum-deposited films are always in a state of tensile stress. Stresses at the substrate–film interface must profoundly affect the observed adhesion.

An entirely different experimental approach is the peel test. This has been used by Derjaguin (1955) in his study of the adhesion of polymers to metal substrates in which, as pointed out in Section IV.A, he deduced that the major part of the work of adhesion was due to the separation of charges in an electrically charged double layer at the interface. Skinner et al. (1953) carried out similar experiments and concluded that only 20% or so of the energy was due to this mechanism. Recently von Harrach and Chapman (1972) carried out some very elegant adhesion experiments in vacuo. They measured the peel strength of metal films evaporated on to an amorphous or crystalline insulator; at the same time the sign and magnitude of the charge remaining on the substrate were determined with a sensitive probe. The results lead to the conclusion that up to about 20% of the stripping energy is due to electrostatic charge separation, 50 to 70% is due to van der Waals interaction, and the remainder to other dissipative processes. Many workers (e.g., Huntsberger, 1967; Gardon, 1967) maintain that peeling tests on metal films are of dubious value, since most of the stripping energy is, in fact, dissipated in plastic deformation of the film. A similar difficulty is involved in the adhesion of polymeric materials, where it is found that hysteresis losses in the polymer can easily account for a large part of the work of dehesion (Andrews and Kinloch, 1973).

More work in this area is highly desirable. A brief summary of some aspects of the field is given by Weaver (1972) and in the monograph edited by Patrick (1968). In addition further attention needs to be given to the more recent work by Schnabel (1969), Krupp (1972), and Derjaguin et al. (1972) showing that adhesion on to a semiconductor surface can be influenced by ultraviolet irradiation because it changes the charge–density of the interfacial electrical double layer.[†]

[†]Many papers on this aspect of adhesion have recently appeared in the monograph "Recent Advances in Adhesion," edited by L. H. Lee (1973).

VI. GENERAL MECHANISM OF ADHESION: SUMMARY

Adhesion between solids arises from the atomic forces exerted across the interface. These forces may be strictly surface forces in the sense that they derive from the surface atoms themselves. For example, surface charges provide truly surface forces; these occur when ionic surfaces are in contact with other ionic solids or with metals (image charges). They will also occur if an electrically charged double layer is formed at the interface. Again, valence bonds provide truly surface forces, though there is some evidence that they need to be thermally activated. Metallic bonds arise partly from surface atoms and partly from the bulk plasmons. All solids will, in addition, experience adhesion due to van der Waals interactions between atoms below the surface layers.

Adhesion interactions may often be calculated, especially for a material adhering to itself, in terms of surface energies. If the free surface energy of the solid is γ, then for every unit area of contact formed between the surfaces, energy of amount 2γ will be released; this may be referred to as the "intrinsic" adhesion. It assumes that the joining process is the exact opposite of separating the body and exposing two units of fresh surface. It does not therefore allow for any relaxation that may occur in the free surface. Even if 2γ provides a reasonable estimate of the interaction energy, there are two factors standing between such a quantity and the observed adhesive *strength* in any adhesion measurement. The first is the general difficulty of deducing strengths from energies. For example, if the force per unit area required to separate two surfaces is σ, it is reasonable to assume that this force operates over a distance x that is not more than a few atomic diameters, say 5 Å. Then in terms of energy

$$\sigma.x = 2\gamma \qquad \text{or} \qquad \sigma = 2\gamma/x.$$

For a van der Waals solid such as paraffin wax, where $\gamma \cong 30\,\mathrm{mJ\,m^{-2}}$, this gives a value for σ of order $12 \times 10^7\,\mathrm{nm^{-2}}$ or over $1000\,\mathrm{kg\,cm^{-2}}$ (the theoretical cleavage strength). The real strength is far lower because of the presence of flaws that produce large stress concentrations. Similar difficulties exist with ductile materials where dislocations play a crucial part in determining the real strength of a solid.

The second difficulty concerns the area of contact. Except for mica, it is almost impossible to obtain molecularly smooth surfaces over appreciable areas. If two surfaces are placed together, real area of contact is usually very much smaller than the geometric area. The interfacial force may itself increase the area of contact as, for example, with a soft, highly elastic material such as rubber. Alternatively, and more usually, the area of contact may be increased by pressing the surfaces together.[†] If the joining load is removed, as it must be if the adhesion is to be measured, there will be a release of elastic stresses in the bulk of the material on either side of the interface. It would appear that the viability of the adhesion junctions when the joining load is removed is determined by the extent to which the junctions can stretch without rupturing. With metals, for example, ductility is the important factor. Ductile materials will generally show strong adhesion. Thus lead and indium adhere strongly at room temperature; metals with higher melting points adhere more strongly at higher temperatures. Again, with clean germanium surfaces the observed adhesion below the brittle–ductile transition temperature (400° C) is small, whereas above this temperature the adhesion can be large.

The behavior of hard solids falls into the same general pattern. With these materials, even though the surface energy and hence the "intrinsic" adhesion may be large, the interface cannot withstand the changes in interfacial shape that occur when the joining load is removed. The observed adhesion may be very small indeed. On the other hand the adhesion of a ductile material to a hard material may be relatively large.

The poor adhesion observed with hard solids is partly due to surface roughness and to the lack of ductility. Another factor may be that, with these materials, bond formation may require an appreciable activation energy. Sapphire and even diamond can be sintered together if the temperature is sufficiently high, but these results on their own would not be sufficient to tell whether these elevated temperatures are required to activate bonds or to provide ductility. However, the experiments showing adhesion between a ductile metal and a hard covalent solid at room temperature and those showing adhesion between two germanium surfaces at temperatures above 400° C suggest that ductility is a more important factor than bond activation.

Adhesion between crystalline solids depends on mutual orientation; between different materials it will depend on lattice mismatch. In general plastic deformation or thermal energy will improve the adhesion of misoriented or mismatched materials. On the other hand, surface films produce a marked reduction in adhesion. In some cases a monolayer is sufficient to reduce the observed adhesion to zero. There are probably two factors involved. First, the film reduces the "intrinsic" adhesion between the surfaces. Secondly, it constitutes a layer of very limited extensibility (or ductility). As the joining load is removed, the weakened junctions are peeled apart by the released elastic stresses themselves. Surface films can be ruptured by produc-

---

[†] The effect of a *tangential* stress in producing junction growth is discussed in Section VII.D.

ing large deformations at the interface. In that case strong adhesions can be obtained.

The behavior of highly elastic solids such as rubber or plastics falls into a separate category. In general they adhere strongly if the surfaces are fairly smooth, in spite of the fact that the interfacial forces are relatively weak. There are probably two reasons for this: since the materials are soft and deformable they easily give a large area of contact, and they can stretch appreciably under the influence of released elastic stresses without rupturing.

Broadly speaking, clean surfaces will adhere to most other clean surfaces. If the observed adhesion is small, it is probably due to some restriction on the area of true contact, to stress concentrations in the interface, or to limited ductility of the adhesional junctions. In addition surface films can greatly weaken the adhesion.

VII. FRICTION

A. Introduction

A review of current work in the field of friction, lubrication, and wear has recently been published (Tabor, 1972) giving over 150 references. The present section will therefore be brief and restricted to those aspects of friction of particular relevance to adhesion.

If two bodies are placed in contact under a normal load W, a finite force is required to initiate or maintain sliding; this is the force of friction and may be designated F_s for static and F_k for kinetic conditions. There are two basic laws of friction: the frictional force is (i) proportional to the load, and (ii) independent of the geometric area of the bodies. These laws are roughly true over a wide range of conditions. There is a third law that states that the kinetic frictional force is about one-third the normal load, but this law is of much more limited validity.

It is now generally accepted (Holm, 1958; Bowden and Tabor, 1950, 1966) that friction between unlubricated surfaces arises from two main factors. The first, and usually the more important factor, is the adhesion that occurs at the regions of real contact; these adhesions, welds, or junctions have to be sheared if sliding is to occur (Ernst and Merchant, 1940). Consequently, if A is the true area of contact and s the average shear strength of the junctions, this part of the friction may be written $F_{adh} = As$. The second factor arises from the ploughing, grooving, or cracking of one surface by asperities on the other. We may call this the deformation term P. Then, if there is negligible interaction between these two processes, we may add them and write

$$F = F_{adh} + F_{def} = As + P. \tag{14}$$

It is clear that the factors of major importance here are the area of real contact A, the strength of adhesion between the surfaces, the shear strength s of the interface, possible interactions between A and s, the deformation component P, and possible interactions between P and the adhesion component of friction.

B. The Deformation Term

If a hard asperity ploughs its way through the surface of a softer material, the ploughing force may be easily calculated if adhesion between the surfaces is negligible. Taking horizontal and vertical and vertical components, it follows that the normal load is supported by the appropriate component of the area of contact, while the horizontal force is supported by the material ahead of the slider. The horizontal force is then equal to the cross-sectional area of the groove multiplied by the contact pressure.

This mechanism was first described by Bowden and Tabor (1943) in their original studies of the mechanism of metallic friction. In its simplest form (Bowden and Tabor, 1964) it leads to the result that for a conical asperity of semiapical angle θ, the coefficient of friction due to ploughing alone is

$$\mu_p = (2/\pi)\cot\theta, \tag{15}$$

so that for a semiapical angle of 60°, $\mu_p \simeq 0.32$; for $\theta = 30°$ (a rather sharp cone) $\mu_p \approx 1.1$. On the other hand, if $\theta = 80°$, $\mu_p \simeq 0.1$.

Since surface roughnesses rarely have slopes greater than a few degrees, the ploughing term will rarely contribute more than obout 0.1 to the observed coefficient of friction. For well-lubricated surfaces this can be an appreciable part of the total friction. But for unlubricated surfaces the contribution from the ploughing or deformation term is small compared with that due to adhesion.

C. The Adhesion Term: The Laws of Friction

With unlubricated surfaces (even if they are not atomically clean) we may ignore the deformation term and attribute the friction to adhesion. In that case we may write

$$F \simeq F_{adh} = As. \tag{16}$$

If the shear strength s of the interface is constant, it is clear that F is proportional to the real area of contact A. Since, as we saw in Section II, this is very nearly proportional to the load and little dependent on the geometry of the surfaces over a wide range of conditions, it is evident that the frictional force will also be roughly proportional to the load and independent of the size of the bodies. This provides the simplest explanation of the two laws of friction described in VII.A. We may write

$$\mu = F/W = As/Ap = s/p, \tag{17}$$

where p is the mean contact pressure over the asperities. Of course, for elastic solids making contact over a single spherical asperity, p is not constant but depends on the load and the radius of the asperity. As may be seen from Eq. (2), in such situations p varies as $W^{1/3}$ so that μ should decrease with increasing load. This is indeed found to be the case with a highly elastic solid such as rubber or, at the other extreme, a very hard material such as carbon fibers. If, however, multiasperity contact occurs, all materials will behave in the way described in Section II, and the coefficient of friction will be little dependent on load or geometry.

D. Clean Metals. The Adhesion Component. Junction Growth

With metals, if the asperities deform plastically, the contact pressure p will be the yield pressure p_0 of the metal. Consequently,

$$\mu = s/p_0. \tag{18}$$

If the surfaces are clean, metallic bonds are formed across the contacting interface. In that case, if the metal does not work-harden appreciably, s is roughly equal to the critical shear stress τ of the metal. The yield pressure p_0 is generally found to be about 5τ (Tabor, 1951); consequently, according to Eq. (18), μ should have a value of about 0.2. In practice most clean metals give enormous coefficients of friction. Even in air they give values of $\mu = 1$ (see Section VII.E). Some part of the discrepancy may be due to the fact that the junction material in the surface layers becomes very highly work-hardened so that s increases more than p, since the latter is determined more by the yield stress of the subsurface material. A far more important factor, particularly for clean ductile metals, is the phenomenon of junction growth under the influences of combined normal and tengential stresses.

Consider a single contact region under a normal load sufficient to produce plastic flow; the contact pressure at the junction will have the value p_0. If a tangential stress s is applied, it will produce further plastic flow, the yield criterion being of the form

$$p^2 + \alpha s^2 = p_0^2, \tag{19}$$

where p is the new normal pressure and α is a constant with a value of about 10 (McFarlane and Tabor, 1950). If $s = 0$, p of course equals p_0. But if s is finite, Eq. (19) can only be satisfied if p diminishes. This is brought about by the surfaces sinking together (see Fig. 9) so that the area of contact is increased with a corresponding drop in p. It should be noticed that initially the major movement is normal to the interface; thus only micro-displacements occur in the tangential direction. This process can continue indefinitely if (i) the surfaces are perfectly clean, and (ii) the metals are very ductile. Junction growth then proceeds until the whole of the geometric area of the surfaces is in intimate contact. This constitutes gross seizure, and it may be virtually impossible to slide the surfaces over one another.

The direct experimental demonstration of junction growth and its role in adhesion was first shown by McFarlane and Tabor (1950) in their study of the behavior of indium in air and by Parker and Hatch (1950). A more critical study was carried out by Bowden and Rowe (1956) in their investigation of the effect of tangential stress on the adhesion of fairly clean metals in a vacuum of 10^{-6} Torr. In these experiments they first pressed their surfaces together with a normal load W and measured the normal force Z required to pull them apart. The ratio Z/W they termed the adhesion coefficient σ by analogy with the definition of the coefficient of friction μ. They then replaced the surface in contact and applied a tangential force F, insufficient to produce gross sliding, and called the ratio $\Phi = F/W$ the tangential prestressing coefficient (when gross sliding occurs, ϕ of course becomes equal to μ). They then removed the tangential stress and measured the adhesion. This was found to have increased as a result of junction growth. A simple theory was derived to allow for the effect of released elastic stresses in breaking peripheral regions of contact. Their experi-

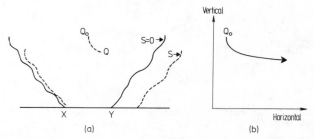

Fig. 9. Behavior of a single asperity making plastic contact with a hard flat surface. (a) Under a normal load alone, contact occurs across XY. When a tangential stress S is applied, the junction grows, as shown by the dotted lines, and the point Q_0 moves to Q. At the initial stages junction growth occurs by the surfaces sinking together. (b) Schematic diagram showing locus of Q_0 as the tangential stress is increased.

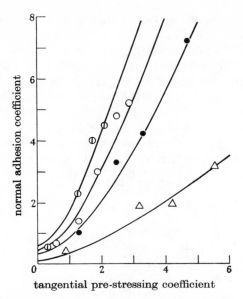

Fig. 10. The influence of a tangential stress (insufficient to produce gross sliding) on the adhesion on clean metals. Experimental results: ∅ gold; ● nickel; ○ platinum; △ silver. The continuous curves are calculated theoretically assuming that the extensibility of an asperity bridge before fracture occurs is of order 100 Å. (From Bowden and Rowe, 1956.)

mental results, showing the effect of the tangential stress in increasing the adhesion coefficient σ, are shown in Fig. 10 together with the theoretical curves. A parallel study has been described by Courtney–Pratt and Eisner (1957) in which electrical resistance measurements were used to follow junction growth as the tangential stress was applied.

These results show that tangential stress can greatly increase the adhesion between ductile materials by increasing the area of real contact. In addition tangential motion, even on a microscale, tends to break up contaminant films. This explains why it is possible to obtain strong adhesion between oxide-covered metals in air by applying a normal pressure together with a tangential force. This was first described by Desaguliers in 1724 in his experiments on lead spheres from each of which he had cut a segment about 1/4 inch in diameter: "I pressed them together with my Hand, with a little Twist, to bring the flat parts to touch as well as I could. The balls stuck so fast..." that a weight of over 16 pounds was required to pull them apart. The experiments by Anderson (1963) and Sikorski (1964) (see Section V.B) involve rather more severe deformation at the interface.

E. Adhesion Component of Friction: Friction of Metals in Air

   In the preceding section we dealt with the situation in which junction growth might proceed indefinitely until the whole of the geometric area is in contact. With real surfaces in air surface contamination brings junction growth to an end before this stage is reached. The reason is that the interface (a mixed fudge of oxide and metal) has a shear strength s_i less than the critical shear stress τ of the metals. Consequently, junction growth continues until the tangential stress in the interface reaches the value s_i; gross sliding then occurs. This will take place (see Tabor, 1959) when

$$p^2 + \alpha s_i^2 = \alpha \tau^2. \tag{20}$$

Putting $s_i = k\tau$ where $k < 1$, sliding occurs when

$$\frac{s_i}{p} = \frac{1}{2^{1/2}(k^{-2}-1)^{1/2}} . \tag{21}$$

The coefficient of friction is

$$\mu = \frac{F}{W} = \frac{A s_i}{A p} = \frac{1}{\alpha^{1/2}(k^{-2}-1)^{1/2}} . \tag{22}$$

The way in which μ varies with the strength of the interface is shown in Fig. 11, where a representative value of $\alpha = 9$ has been taken. It is seen that for $k = 1, \mu = \infty$; this corresponds to gross seizure. For an interface with shear strength about 95% of $\tau(k = 0.95)$ sliding occurs for a value of $\mu \simeq 1$. Finally, for very small values of k (the situation of sliding takes place in a thin soft surface film of metal or lubricant)

$$\mu = \frac{s_i}{p} \approx \frac{s_i}{p_0} = \frac{\text{shear strength of interface}}{\text{hardness of metal}} . \tag{23}$$

Here there is very little junction growth; thus the area of contact is essentially that of the static contact (Tabor, 1959).

F. Friction of Metals and Alloys: Effect of Structure

   Buckley and Johnson (1966) have carried out friction and wear experiments on a variety of alloys in a vacuum of 10^{-10} Torr after baking out at $200°$ C. The surfaces are repeatedly slid over one another for periods of the order of 1 hour, and it is found that certain alloys and metals give relatively low friction and wear even under these severe conditions. For example, a titanium alloy containing 21% aluminum sliding on a stainless-steel surface gives a friction coefficient μ of only about 0.4; with pure titanium sliding

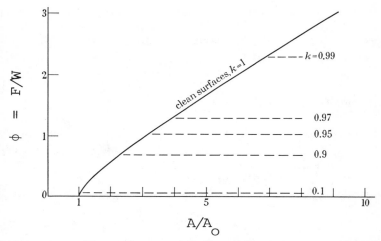

Fig. 11. Junction growth and coefficient of friction for different values of the interfacial shear strength s_i, where s_i is expressed in terms of the critical shear strength τ of the metal ($s_i = k\tau$). As the tangential force coefficient $\varnothing$ ($= F/W$) is increased, junction growth proceeds until at some critical stage determined by the value of k, gross slip occurs. For $k = 1$ junction growth proceeds indefinitely until the surfaces have seized together over the whole of their available area. For $k = 0.95$ slip occurs when the area of contact has grown about threefold and the force coefficient (now equal to the coefficient of friction) has a value of about 1. For a small value of k (the situation when sliding takes place in a thin film of a soft metal or a lubricant) there is very little junction growth.

on the same surface there is gross seizure ($\mu \gg 1.6$). The authors suggest that this is connected with the structure of the metal or alloy and particularly with a favorable ratio of the a to c dimension in hexagonal structures, a large $c:a$ ratio giving low friction and wear. A different approach is due to Bowden and Childs (1969). They studied the friction of various metals in a high vacuum and were able to change the mechanical properties of the surfaces by operating over a range of temperature from 20° to 300° K. Their results suggested that the friction and wear of clean metals are heaviest if the materials are ductile; the sliding interaction becomes less severe as the ductility is reduced. This may be associated with an increase in work-hardening rate at low temperatures, a ductile-brittle transition over some critical temperature range, or a phase change. For example, Powell (1969) found such an effect with Co, which is cubic above 600° K and hexagonal below and shows a marked reduction in friction and wear below this temperature. These results agree with those of Buckley and Johnson but provide a broader generalized explanation.

Metals and alloys may show appreciable frictional anisotropy. This is shown most simply by studying the friction of hard sliders over the surface of a single crystal. In certain directions a hill of work-hardened material forms ahead of the slider and causes an increase in friction. This is the "bourrelet frontal" investigated so elegantly by Courtel and his colleagues (Courtel, 1965).

G. Friction of Brittle Solids

With brittle solids the frictional behavior resembles that of metals for two basic reasons. First, there is normally strong interfacial adhesion; secondly, the hydrostatic stresses around the regions of contact inhibit brittle fracture and although some cracking may occur the deformation is dominantly plastic. There may even be some junction growth. However, there is a limit to this. As the area of contact increases, the normal pressure falls off and the ability of the hydrostatic stress to inhibit brittle failure diminishes. Junction growth comes to an end as the material ceases to be ductile. It is for this reason that, even with very clean surfaces, μ rarely exceeds a value of ~ 1.

Brittle crystalline solids can show marked frictional anisotropy, if, for example, a hard slider traverses the surface of a single crystal. When the deformation is gentle, the behavior resembles that of ductile materials and metals, the higher friction being associated with the piling up of a hill of work-hardened material *ahead* of the slider. When the deformation is predominantly brittle, frictional anisotropy again occurs and again is largely due to an increase in the ploughing term. However, the high friction observed in certain crystallographic directions is due not so much to a piling up of material as to an increase in penetration by the slider into the brittle crystal (Bowden and Brookes, 1966).

The ploughing mechanism may account, in part, for the frictional anisotropy observed when a diamond stylus slides over a diamond surface in air. Recently Wilks and Wilks (1972) have revived the ideas of Tolkowsky (1920) and suggested that the frictional anisotropy of diamond is due to crystallographic roughnesses, but they do not clearly elucidate the dissipative process. In addition, although in their view roughness critically affects the wear properties in the way described by Tolkowsky, damage to the surface does not contribute to the frictional resistance. These results may be contrasted with the experiments of Bowden and Hanwell (1964), which show that if diamond is slid over diamond in a moderate vacuum, the friction gradually increases, and this increase is accompanied by marked fragmentation of the diamond surfaces. These effects are clearly due to an increase in the strength of the interface. If a small amount of air is admitted, the friction drops but again rises after repeated traversals.

We have already referred to the poor normal adhesion of clean diamond surfaces in a very high vacuum. It is possible that the cleaning methods adopted leave some residual surface film or graphitized layer. However, it is more likely that the poor normal adhesion is due to the brittleness of diamond.

H. Friction of Lamellar Solids

It is now generally agreed that the low friction of polycrystalline graphite is associated with the presence of adsorbed films. When these films are removed, adhesion is strengthened, the friction rises, and there is an enormous increase in wear (Savage, 1948). There is still some dispute as to whether the cleavage face in graphite is intrinsically a low-energy surface. Some experiments by Bryant et al. (1964) suggest that this is not so, which would support the view of Rowe (1960) that the friction of graphite is due to the fracture or cleavage of crystals and that adsorbed gases penetrate to some depth into the lattice and thus reduce the cleavage strength of the graphite crystallites. Most other workers consider that the cleavage face is a low-energy surface and that friction involves the sliding of the crystallites themselves over one another. The edges are considered to be the high-energy sites at which adsorption of gases and vapors readily occurs (Deacon and Goodman, 1958). On this view any interaction involving edges gives a high friction when the edges are clean, a lower friction when the edges are covered with adsorbed films.

The anisotropic frictional properties of single crystals of natural and synthetic graphite have recently been investigated by Skinner et al. (1973) in a scanning electron microscope. No special precautions were taken to clean the surfaces. Since natural graphite crystals are extremely thin, the edge planes studied were those of pyrolytic graphite. The loads used were very small (1 to 400 mg), and the slider consisted of a fine tungsten stylus. On the edge planes the friction was of order $\mu \simeq 0.3$. On the basal plane, if the load was below 40 mg and if no surface damage occurred, the friction was extremely low, often less than $\mu = 0.01$. If, however, penetration of the basal plane occurred or if peeling took place, the friction rose to values of the order $\mu = 0.2$ or more. Evidently interaction with the cleavage planes involves very weak interfacial forces, with the edge planes much larger forces.

With MoS_2 it is now generally agreed that the low friction is essentially a property of the intrinsic structure of the material, MoS_2 being extremely weak along the shear planes (for a survey see Winer, 1967). The presence of adsorbed vapor and the other impurities tends, if anything, to increase the friction (Haltner, 1964). On the other hand, intercalated long chain organic molecules reduce the interplanar bonding and friction is reportedly reduced.

One of the more interesting developments in this field is the effect of crystal structure on the frictional properties of lamellar solids of the MoS_2 type (transition-metal dichalcogenides). Some evidence suggests, as with the hexagonal metals and alloys studied by Buckley and Johnson, that a large $c:a$ ratio is advantageous (Salomon, 1966). Experiments by Giltrow (1966), however, suggest that the effectiveness of lamellar compounds depends mainly on the extent to which they are stoichiometric, the more perfect their structure the lower the friction. A somewhat different point of emphasis has been made by Bowden et al. (1968). They stress the importance of inter-lamellar bonding. For example, with titanium sulphide interstitial titanium atoms can be incorporated within one lamella to give strong bonding with a neighboring lamella. This lamellar sulphide has in fact poor antifriction properties. On general grounds one would expect that intercalated compounds that weaken the interlamellar bonding would also reduce the friction, but detailed evidence on this is lacking.

I. Friction of Rubbers and Polymers

For completeness we may add a brief account of the friction of rubbers and polymers. It is again found that the friction is largely due to interfacial adhesion and to a lesser extent to the ploughing or deformation of the elastomer by asperities on the other surface (Grosch, 1963; Ludema and Tabor, 1966). The adhesion is essentially due to van der Waals interactions. The main difference between the frictional behavior of these materials and those so far discussed is that, as they are viscoelastic, their strength properties are markedly dependent on temperature and rate of deformation. This is reflected in their frictional behavior; the friction depends on temperature and speed (Vinogradov et al., 1970). A molecular theory of rubber friction has been described by Bartenev (1954, 1973) and Schallamach (1963). Other theories have been described by Bulgin et al. (1962), Savkoor (1965), Kummer and Meyer (1966) and are summarized in a recent book by Moore (1972). However, some later observations by Schallamach (1971) suggest that these theories do not adequately describe the detailed behavior at the sliding interface.

The low friction of PTFE (Teflon) arises from two factors: the weak adhesion between the molecular chains, and the fact that, at the sliding interface, the material develops a preferred orientation with the chains extended parallel to the direction of sliding. Low density polyethylene with its straggly side groups is unable to form such an interface, and its friction is relatively high. On the other hand, high density polyethylene, in which such groups are practically nonexistent, shows a frictional behavior closely resembling that of PTFE (Pooley and Tabor, 1972).

J. Adhesion in Friction: Friction under a Negative Load

An interesting example illustrating the role of adhesion in friction is shown by experiments in which finite frictional forces are observed even when the surfaces are under a negative load, i.e., being pulled apart. The first clear description of this effect is due to Lazarov and Denikin (see Ahkmatov, 1966). In their experiments two "flat" smooth metal surfaces were placed in contact with a film of stearic acid between. (Stearic acid is recognized as a very effective boundary lubricant.) The thickness of the film (40 monolayers, about 10^{-5} cm) was larger than the surface irregularities, so metallic contact through the film did not occur. A finite force was required to separate the metal surfaces. For any pulling force less than this a finite shear force was required to initiate sliding. The version of this work in Akhmatov's monograph (1966) does not provide enough information for useful calculations to be made, but a recent article in Russian by Akhmatov and Uchuvatkin (1973) provides calculations of the forces in and across the film due to van der Waals interactions. Although these observations are of great interest, it is not clear in any of these papers whether the surfaces remain in contact under a negative load once sliding has commenced.

Akhmatov's experiments represent the behavior of surfaces acting under lubricated conditions. A more direct study for unlubricated surfaces is that due to Skinner and Gane (1972). They studied the friction of surfaces at very small loads in a scanning electron microscope, using a delicate galvanometer movement as the loading device. For gold sliding over a smooth diamond surface there was negligible transfer of metal to the diamond and a "flat" slowly formed on the gold slider partly as the result of creep, partly due to the action of the normal and tangential stresses. For an initial normal load of $690 \, \mu N$ (6.9×10^{-2} g) it was found that the worn flat had a diameter of $\sim 6 \times 10^{-3}$ mm, i.e., an area of about $0.25 \times 10^{-4} \, mm^2$. Thus the nominal pressure over the worn contact region under the normal load was of order $3 \, kg \, mm^{-2}$. This is considerably less than the yield pressure of gold particularly since the size effect (see Section II.E) would tend to impart to the gold a yield stress appreciably higher than its bulk value. The contact was therefore largely elastic. At this stage there was a finite adhesion between the slider and the diamond surface. The force to pull the surfaces apart was about $100 \, \mu N (10^{-2} \, g)$. The surfaces could be set sliding over one another and the steady kinetic friction measured. As the normal load was reduced from its initial value, the tangential force decreased but remained finite even for negative loads, if these were less than the adhesive force. A typical set of results is given in Fig. 12 for both gold and lead sliders. The most likely explanation is that true contact occurs over a very small fraction of the "flat." (See Fig. 13.) The attraction occurs mainly from the relatively large

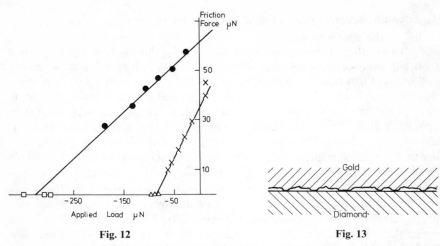

Fig. 12

Fig. 13

Fig. 12. Steady state sliding under a negative load for gold and lead styli on diamond. Lead, preload 270 μN : ○ friction; □ static adhesion. Gold, preload 690 μN: × friction; △ static adhesion.

Fig. 13. Proposed contour between worn gold slider and diamond surface. It is suggested that the near-contact areas provide the major part of the observed adhesional force.

area of near-contact where the surfaces are probably less than 50 Å apart. A simple calculation, assuming van der Waals forces, would account for the observed adhesion if the average separation was of order 10–15 Å. This force acts in conjunction with any applied load to increase the true area of contact and so give a fairly linear dependence of friction on load. It also produces a finite area and a finite frictional force for a negative load if this is insufficient to overcome the attractive force. One may ask if energy is also dissipated by the sliding of those atoms in the gold, which are at a distance of 10–15 Å from the atoms in the diamond. It would seem that at this separation the potential energy profile is smooth and does not reflect the atomic graininess of the surfaces; consequently, the sliding of these atoms will not be a dissipative process. By contrast at the real contact areas there is the short-range hill-and-valley profile to the surface potential, and this involves energy dissipation when sliding occurs.

The results show that the frictional force may be written

$$F = \mu[W + Z], \tag{24}$$

where W is the applied load and Z the adhesive force. This equation was first derived by Derjaguin in 1934 from a somewhat different friction model. In his analysis the area of contact remains constant and the effect of the adhesive forces is to increase the force of interaction between the interfacial atoms. In the above experiments, by contrast, it is assumed that the effect of the adhesive force is to increase the area of real atomic contact where the frictional process actually takes place. We discuss this in what follows in a rather simplified form.

K. Adhesion in Friction: Effect of Normal Pressure

The adhesional theory of friction states that the frictional force may be written as

$$F = As, \tag{25}$$

where s is the shear strength per unit area of the interface. In a theoretical discussion of the shear properties of a crystalline solid, Derjaguin (1934) suggested that the ideal stress to produce sliding of one set of planes over another may be estimated by regarding the atomic profile as equivalent to atomic roughnesses of average slope θ. Thus if the attractive forces acting normally across the pair of planes is Z per unit area, the tangential force to produce sliding will be $Z \tan \theta$, which may be written as μZ, where $\mu = \tan \theta$. If now an external normal load W is applied, the force to produce sliding will be $(Z + W)\tan \theta = (W + Z)\mu$. The frictional force may therefore be written in the form

$$F = \mu(W + Z). \tag{26}$$

As pointed out above this does not take into account the way in which the area of contact itself is influenced by W and/or Z. Further, this original theory was derived by Derjaguin several years before it was realized that the true strength of real crystals is determined by dislocations and other imperfections. It seems more realistic therefore to revert to the simple view that the bulk shear strength of a solid may increase with pressure; thus we may write[†]

$$S = S_0 + \alpha p, \tag{27}$$

where S_0 is the shear strength when the external applied load is zero and so corresponds to μZ; while the term αp corresponds to the term μW. To carry this forward we must now bring in the area of contact A. Since $A = W/p$, we have

$$F = As = W/p(s_0 + \alpha p) = Ws_0/p + W\alpha = W\mu_0 + W\alpha', \tag{28}$$

[†] In his study of the friction of polymers Adams (1963) suggested that s is a weak power function of the pressure, i.e., $s = s_0 p^\beta$, where β is constant of order 0.2.

where μ_0 is the coefficient of friction if pressure had no effect on the shear strength. This equation resembles Eq. (26), but there the second term was due to intermolecular attraction. Soviet physicists have, in the past, been greatly fascinated by a "two-term" law of friction, following Coulomb's classical paper that first proposed it. [For a brief review see the English translation of Akhmatov's interesting monograph (1966) and the more technical book by Kragelsky (1965).] It is evident that various models can lead to such equations, although the physical significance of these terms may differ. Indeed as Kragelsky (1971) pointed out in a recent review article, the number of terms in a friction equation really reflects the number of important processes involved in the friction. These may be adhesive and ploughing (two), adhesion and molecular attraction (two), adhesion and pressure dependence of s (two), adhesion, pressure dependence of s, and ploughing (three), etc. Indeed it is not difficult to derive, on the basis of our preceeding discussion, a friction equation with even more terms. For example, if the normal load is W and the adhesion force is Z, the area of contact will be a function of W and Z; the simplest form will be $(W + Z)/p$ where p is an appropriate modulus that may include geometric factors and elastic and plastic properties. The shear strength will be a function of the contact pressure; the simplest relation is $s = s_0 + \alpha p$. The friction equation then becomes

$$F = [(W + Z)/p][s_0 + \alpha p] + F_{\text{ploughing}}.$$

We now have a five-term law of friction, but how far it is meaningful depends on the situation to which it is applied.

The main defect of Eqs. (26), (27), and (28) is not in their derivation but in their limited applicability. For clean crystalline solids in contact (as first discussed by Derjaguin) the shearing of an interface cannot be expressed in terms of the shear stress alone; yielding involves the yield criterion of the solid and the *combined* effects of the normal and tangential stress. If these equations are at all meaningful they can only be applied to systems in which solids are separated by weak films in which the shearing occurs. If the shear strength of these materials depends on pressure, the equations will both describe the observed behavior and be physically realistic. Such an effect has recently been described by Briscoe *et al.* (1973).

VIII. GENERAL MECHANISM OF FRICTION

The friction between unlubricated solids is due to two main factors. The first is the adhesion occurring at the regions of real contact. The second arises if the asperities on a hard surface produce grooving or cutting in the other. With metals the ploughing action involves plastic displacement, with brittle solids some cracking or fragmentation, with rubbers and polymers hysteresis losses. Other factors, some of which have been discussed in previous sections, may also play a part in the frictional process, but except in special circumstances these are generally of only tertiary significance.

In this chapter we have emphasized the role of adhesion and the important part played by the deformation characteristics of the solids. Solids that deform easily and isotropically will tend to give very high coefficients of friction. On the other hand, brittle solids will give restricted junction growth and lower coefficients of friction. Again, solids with well-defined anisotropy may show anisotropic frictional properties. This is most marked with lamellar solids that are often low friction materials, though this is by no means always the case. With viscoelastic solids the deformation properties are rate- and temperature-dependent, and there is a corresponding variation of friction with sliding speed and temperature.

Although deformation properties are important, the factor that has the largest overall influence on friction is the cleanliness of the surfaces. A small trace of oxygen or some other contaminant can radically reduce the adhesion and friction. It is probable that in practical affairs involving the sliding of unlubricated metals, atmospheric contamination is the main factor responsible for reducing friction and wear and for preventing total seizure. With polymers and brittle materials the situation is not as clearly defined, but, in general, surface contamination has less effect than with metals.

ACKNOWLEDGMENTS

I wish to thank Dr. B. J. Briscoe and Dr. N. Gane for critical and constructive comments.

REFERENCES

ADAMS, N. (1963). *J. Appl. Polymer Sci.* **7**, 2075–2103.
AKHMATOV, A. S. (1966). Molecular Physics of Boundary Lubrication. Israel Programme for Scientific Translations, Jerusalem.
AKHMATOV, A. S., and UCHUVATKIN, G. N. (1973). *In* "Electrical Phenomena in the Friction, Cutting and Lubrication of Solids" (M. M. Kruschov and V. A. Bobrovskii, eds.), pp. 7–12. Nauka, Moscow (In Russian).
ANDERSON, D. L. (1963). *Wear* **3**, 253–273.
ANDERSON, D. L., CHRISTENSEN, H., and ANDREATCH, P. (1957). *J. Apol. Phys.* **28**, 293.
ANDREWS, E. H., and KINLOCH, A. J. (1973). *Proc. Roy. Soc.* (*London*) **A332**, 385–399, 401–414.
ARCHARD, J. F. (1957). *Proc. Roy. Soc.* **A243**, 190–205.
ARCHARD, J. F. (1974). *Tribology Int.* (*Oct.*), 213–220.

526   D. TABOR

BAILEY, A. I., and COURTNEY-PRATT, J. S. (1955). *Proc. Roy. Soc.* (*London*) **A227**, 500–515.

BAILEY, A. I., and DANIELS, H. (1972). *Nature Phys. Sci.* **240**, 62–63.

BAILEY, A. I., and KAY, S. M. (1967). *Proc. Roy. Soc.* (*London*) 47–56.

BARDEEN, J. (1949). *Phys. Rev.* **71**, 717.

BARDEEN, S.1964). *In* "Solid Surfaces" (H. Gates, ed.), North Holland Publ. Co., Amsterdam, p. 381.

BARTENEV, G. M. (1954). *Dokl. Akad. Nauk SSSR* **96**, 1161–1164.

BARTENEV, G. M. (1973). "Trenye in Iznaahivanya Polimerov."

BELSER, R. B., and HICKLIN, W. (1956). *Rev. Sci. Instrum.* **27**, 293–296.

BENJAMIN, P., and WEAVER, C. (1960). *Proc. Roy. Soc.* (*London*) **A254**, 177–83.

BENJAMIN, P., and WEAVER, C. (1961). *Proc. Roy. Soc.* (*London*) **A261**, 516–531.

BENJAMIN, P., and WEAVER, C. (1963). *Proc. Roy. Soc.* (*London*) **A274**, 267.

BICKERMANN, J. J. (1968). "The Science of Adhesive Joints," 2nd ed. Academic Press, New York.

BOWDEN, F. P., and BROOKES, C. A. (1966). *Proc. Roy. Soc.* (*London*) **A295**, 244–258.

BOWDEN. F. P., and CHILDS, T. H. C. (1969). *Proc. Roy. Soc.* (*London*) **A312**, 451–466.

BOWDEN, F. P., and HANWELL, A. E. (1964). *Nature* **201**, 1279–1281.

BOWDEN, F. P., and ROWE, G. W. (1955). *Proc. Roy. Soc.* (*London*) **A228**, 1–9.

BOWDEN, F. P., and ROWE, G. W. (1956). *Proc. Roy. Soc.* (*London*) **A233**, 429–442.

BOWDEN, F. P., and TABOR, D. (1943). *J. Appl. Phys.* **14**, 141–151.

BOWDEN, F. P., and TABOR, D., (1950). "Friction and Lubrication of Solids," Pt. I. Clarendon Press, Oxford.

BOWDEN, F. P., and TABOR, D. (1964). "Friction and Lubrication of Solids," Pt. II. Clarendon Press, Oxford.

BOWDEN, F. P., and TABOR D. (1966). *Brit. J. Appl. Phys.* **17**.

BOWDEN, F. P., and THROSSEL, W. R. (1957). *Proc. Roy. Soc.* (*London*) **A209**, 297.

BOWDEN, F. P., GREENWOOD, J. H., and IMAI, M. (1968). *Proc. Roy. Soc.* (*London*) **A304**, 157–169.

BRISCOE, B. J., SCRUTON, B., and WILLIS, R. F. (1973). *Proc. Roy. Soc.* (*London*) **A333**, 99–114.

BRYANT, P., GUTSHALL, P. L., and TAYLOR, L. H. (1964). *Wear* **7**, 118–126.

BUCKLEY, D. H. (1969). *J. Adhes.* **1**, 264.

BUCKLEY, D. H. (1971). NASA SP–277.

BUCKLEY, D. H. (1972). *Wear* **20**, 89.

BUCKLEY, D. H., and JOHNSON, R. L. (1966). NASA Rept. TND-3235.

BULGIN, D., HUBBARD, G. D., and WALTERS, M. H. (1962). *Proc. 4th Rubber Technology Conf.*, London, pp. 173–188.

BUTLER, D. W., STOODART, C. T. H., and STUART, P. R. (1970). *J. Phys. D.: Appl. Phys.* **3**, 877.

COHEN, S. C., and TABOR, D. (1966). *Proc. Roy. Soc.* (*London*) **A291**, 186–207.

COURTEL, R. (1965). *Met. Corrosion-Ind.* **473**, 1–6.

COURTNEY–PRATT, J. S., and EISNER, E. (1957). *Proc. Roy. Soc.* (*London*) **A238**, 550.

CZICHOS, H. (1972). *J. Phys. D.* **5**, 1890–97.

DAVIS, D. K. (1973). *J. Phys. D.: Appl. Phys.* **6**, 1017–24.

DAYSON, C., and LOWE, J. (1972). *Wear* **21**, 263–288.

DEACON, R. F., and GOODMAN, J. F. (1958). *Proc. Roy. Soc.* **A243**, 464–482.

DENISOV, P. V. (1955). Quoted in Akhmatov's Monograph, q.v.

DERJAGUIN, B. V. (1934). *Zh. Fiz. Khim.* **5**, 1165.

DERJAGUIN, B. V. (1955). *Research* (*London*) **8**, 70.

DERJAGUIN, B. V., TOPOROV JuP., ALEINIKOVA, I. N., and BURTA-GAPONOVITCH, L. N. (1972). *J. Adhesion* **4**. 65–71.

DESAGULIERS, J. T. (1724). *Phil. Trans. Roy. Soc.* (*London*) **33**, 345.

DJAFARI-ROUHANI, M., and SCHUTTLER, R. (1973). *Surface Sci.* **38**, 503.

DYSON, J., and HIRST, W. (1954). *Proc. Phys. Soc.* (*London*) **B67**, 309–312.

EASTERLING, K. E., and SWANN, P. R. (1971). *Acta Met.* **19**, 117.
ELEY, D. D. (ed.) (1962). "Adhesion" Oxford Univ. Press, London and New York.
ERNST, H., and MERCHANT, M. E. (1940). *Conf. Friction and Surface Finish*, pp. 76–101. M.I.T.
FERRANTE, J., and SMITH, J. R. (1973). *Surface Sci.* **38**, 77–92.
GANE, N. (1970). *Proc. Roy. Soc.* **A317**, 367–391.
GANE, N. (1973). Personal communication
GANE, N., PFAELZER, P. F., and TABOR, D. (1974). *Proc. Roy. Soc. (London)* **A340**, 495–517.
GARDON, J. L. (1967). *In* "Treatise on Adhesion and Adhesives" (R. Patrick, ed.), *Vol I*, pp. 269–324. Dekker, New York.
GILTROW, J. P. (1966). R.A.E. Rep. No. 66184.
GREENWOOD, J. A., and WILLIAMSON B. (1966). *Proc. Roy. Soc.* **A295**, 300–319.
GROSCH, K. A. (1963). *Proc. Roy. Soc.* **A274**, 21–39.
HALTNER, A. J. (1964). *Wear* **7**, 102–117.
HAMAKER, H. C. (1937), *Physica* **4**, 1058.
HEAVENS, O. S. (1950). *J. Phys. Radium* **11**, 355.
HENZLER, M. (1973). *Surface Sci.* **36**, 109–122.
HOLM, R. (1958). "Electrical Contracts". Springer, Berlin.
HOUWINK, R., and SALOMON, G. (1967). "Adhesion and Adhesives" Elsevier, Amsterdam.
HUNTSBERGER, J. R. (1967), *In* "Treatise on Adhesion and Adhesives" (R. Patrick, ed.), Vol. I, pp. 119–149. Dekker, New York.
ISRAELACHVILLI, J. N., and TABOR, D. (1973). *Proc. Roy. Soc. (London)* **A331**, 19–38.
ISRAELACHVILLI, J. N., and TABOR, D. (1973). *In* "Progress in Surface and Membrane Science" (J. F. Danielli, M. D. Rosenberg, D. A. Cadenhead, eds.). Academic Press, New York.
JOHNSON, K. L., KENDALL, K., and ROBERTS, A. D. (1971). *Proc. Roy. Soc.* **A324**, 301–313.
KELLER, D. V. (1963). *Wear* **6**, 353–365.
KELLER, D. V. (1967). *J. Appl. Phys.* **38**, 1896.
KELLER, D. V. (1972). *J. Vac. Sci. Technol.* **9**, 133.
KENDALL, K. (1971). *J. Phys. D. (Appl. Phys. London)* **4**, 1186–1195.
KENDALL, K., and TABOR, D. (1971). *Proc. Roy. Soc. (London)* **A323**, 321–340.
KING, R. F., and TABOR, D. (1956). *Proc. Roy. Soc. (London)* **A236**, 250–264.
KRAGELSKY, I. V. (1965). "Friction and Wear." Butterworths, London and Washington, D.C.
KRAGELSKY, I. V. (1971). *Conf. Nature Friction Solids, Minsk, 1970*, pp. 262–280. Nauka i Technika, Minsk (In Russian).
KRAGELSKY, I. V., and SABELNIKOV, V. P. (1957). *Inst. Mech. Eng. Conf. Lubrication Wear*, pp. 247–251. Inst. Mech. Eng. London.
KRAGELSKY, I. V., BELY, V. A., and SVIRIDYONOK, A. I. (1974). *Int. Symp. Advan. Polym. Friction Wear*, Amer. Chem. Soc., Los Angeles, California.
KRUPP, H. (1967). *Advan. Colloid Interface Sci.* **I**, 111.
KRUPP, H. (1972). *J. Adhesion* **4**, 83–86.
KUMMER, H. W., and MEYER, W. E. (1966). *J. Mater.* **1**, 667.
LAZAREV (1928). Quoted in Akhmatov's Monograph, q.v.
LEE, L. H. (ed.) (1973). "Recent Advances in Adhesion." Gordon and Breach, New York.
LIFSHITZ, E. M. (1956). *Sov. Phys.—JETP* **2**, 73.
LIN, D. S. (1971). *J. Phys. D. (London)* **4**, 1977–1990.
LODGE, A. S., and HOWELL, H. G. (1954). *Proc. Phys. Soc.* **B67**, 89.
LONDON, F. (1930). *Z. Phys.* **63**, 245.
LUDEMA, K. C., and TABOR, D. (1966). *Wear* **9**, 329–348.
MAHANTY, J., and NINHAM, B. W. (1973). *J. Phys. C.: Solid State Phys. (London)* (submitted).
MCFARLANE, J. S., and TABOR, D. (1950). *Proc. Roy. Soc. (London)* **A202**, 244–253.
MELMED, A. J., and CARROLL J. J. (1973). *J. Vac. Sci. Technol.* **10**, 164.

METSIK, M. S. (1973). *In* "Recent Advances in Adhesion" (L. H. Lee, ed.), p. 145. Gordon and Breach, New York.

MØLGAARD, J. (1962). *Proc. Phys. Soc. (London)* **79**, 516–534.

MOORE, D. F. (1972). "The Friction and Lubrication of Elastomers," Pergamon, Oxford.

NAVARA, E., and EASTERLING, K. E. (1971). *Int. J. Powder Met.* **7**, 11.

OSIAS, J. R., and TRIPP, J. H. (1966). *Wear* **9**, 388.

PARENT, A. (1704). *Mémoires de l'Académie Royale*, 173.

PARKER, R. C., and HATCH, D. (1950). *Proc. Phys. Soc. (London)* **B63**, 185.

PATRICK, R. (ed.) (1968). "Treatise on Adhesion and Adhesives," Dekker, New York.

PFAELZER, P. (1971). Ph.D. Thesis, Univ. of Cambridge.

PFEFFERKORN, G. (1953). *Naturwissenschaft* **40**, 551.

PHILLIPS (1973). Private communication.

POOLEY, C. M., and TABOR, D. (1972). *Proc. Roy. Soc.* **A329**, 251–274.

POWELL, B. D. (1969). Ph.D. Thesis, Univ. of Cambridge.

RARATY, L. E., and TABOR, D. (1958). *Proc. Roy. Soc.* **A245**, 184–201.

ROBERTS, A. D., and TABOR, D. (1971). *Proc. Roy. Soc. (London)* **A325**, 323–345.

ROWE, G. W. (1960). *Wear* **3**, 274–285.

SALOMON, G. (1966). *T.N.O. News* **21**, 39.

SAVAGE, R. H. (1948). *J. Appl. Phys.* **19**, 1–10.

SAVKOOR, A. R. (1965). *Wear* **8**, 222–237.

SCHALLAMACH, A. (1963). *Wear* **6**, 375–382.

SCHALLAMACH, A. (1971). *Wear* **17**, 301–312.

SEMENOFF, A. P. (1958). "Skhvativanye Metallov." Masngiz, Moscow.

SEMENOFF, A. P. (1961). *Wear* **4**, 1–9.

SCHMIT, J., and LUCAS, A. A. (1972a), *Solid State Commun.* **11**, 415–418.

SCHMIT, J., and LUCAS, A. A. (1972b). *Solid State Commun.* **11**, 419–423.

SCHNABEL, W. (1969). Lichtmodulierte Elektrostatische Doppelschicktheftung. Dissertation, Karlsruhe. Quoted by H. Krupp (1972), q.v.

SIKORSKI, M. E. (1964). *Wear* **7**, 144–162.

SKINNER, J., and GANE, N. (1972). *J. Phys. D: Appl. Phys.* **5**, 2087–2094.

SKINNER, J., and GANE, N. (1973). *Wear* **25**, 381–384.

SKINNER, J., GANE, N., and TABOR, D. (1971). *Nature (London)* **232**, 195–6.

SKINNER, S. M., SAVAGE, R. L., and RUTZLER. (1953). *J. Appl. Phys.* **24**, 438–450.

TABOR, D. (1951). "The Hardness of Metals." Oxford Univ. Press (Clarendon), London and New York.

TABOR, D. (1959). *Proc. Roy. Soc. (London)* **A251**, 378–393.

TABOR, D. (1972). *Surface Colloid Sci.* **5**, 245–312.

TABOR, D. (1975). *Wear* **32**, 269–271.

TAKAJI, R. (1957). "Electron Microscopy" (*Proc. Regional Conf. Asia, Oceana, 1st, Tokyo, 1956*) p. 297.

TOLKOWSKY, M. (1920). D.Sc. Thesis, London.

TREFILOV, V. I., and MIL'MAN YU. V. (1964). *Sov. Phys.-Dokl.* **8**, 1240.

VINOGRADOV, G. W., BARTENEV, G. M., ELKIN, A. I., YANOVSKY, Y. G., NICKOLAYEV, V. N., and FRENKIN, E. I. (1969). *Brit. J. Phys.* **2**, 1687.

VINOGRADOV, G. W., BARTENEV, G. M., ELKIN, A. I., and MIKHAYLOV, V. K. (1970). *Wear* **16**, 213–219.

VON HARRACH, H., and CHAPMAN, B. N. (1972). *Proc. Int. Conf. Thin Films, Venice* **II**, 157–161.

VOYUTSKII, S. S. (1963). "Autoadhesion and Adhesion of High Polymers." Wiley, New York.

WEAVER, C. (1972). Faraday Soc. Spec. Discuss., "Adhesion of Metals to Polymers. No. 2, Nottingham, U.K.

WEISS, P. (ed.) (1962). "Adhesion and Cohesion." Elsevier, Amsterdam.
WHITEHOUSE, D. J., and ARCHARD, J. F. (1970). *Proc. Roy. Soc.* **A316**, 97–121.
WILKS, E. M., and WILKS, J. (1972). *J. Phys. D: Appl. Phys.* (*London*) **5**, 1902–1919.
WILLIAMSON, J. B. P. (1968). Interdisciplinary Approach to Friction and Wear. *NASA Symp. SP-181, San Antonio, 1967,* pp. 85–142.
WILLIAMSON, J. B. P., and HUNT, R. T. (1972). *Proc. Roy. Soc.* (*London*) **A327**, 147–157.
WINER, W. O. (1967). *Wear* **10**, 422–452.
YOUNG, R. (1972). *Rev. Sci. Instrum.* **43**, (7), 999.
ZISMAN, W. A. (1963). *Ind. Eng. Chem.* **55**, 18.

Index for Volume II

A

Abrasion and polishing, 479
Adatoms, 282
Adhesion, experimental study of
  covalent solids, 503
  ionic solids, 499
  metals, 495-499
  methods, 494
  role of ductility in, 506
  thin film, 507
  van der Waals solids, 499-503
Adhesion in friction, role of, 521
  effect of normal pressure on, 523
Adhesion, theoretical nature of
  covalent, 493
  insulators, 491, 508
  ionic, 491
  metallic, 489
  van der Waals, 492
Adsorption, on solid surfaces, 477
AEAPS, *see* Auger electron appearance
    potential spectroscopy
AES, *see* Auger electron spectroscopy
Ag, 455, 464
Alkali chemisorption, 366-367
Alkali halides, surface diffusion of, 327
Alloys, surface diffusion of, 325
Aluminum, adhesion of, 508
Anderson model, 341, 343, 346, 360,
  368-369
Anharmonicity, 420, 423, 443-448, 457,
  459, 460
Appearance potential spectroscopy, *see* Soft
    x-ray appearance potential spectroscopy

APS, *see* Soft x-ray appearance potential
  spectroscopy
Area of contact
  in adhesion studies, 497
  between surfaces, analytical models, 481
  true, experimental studies of, 487-489
Atomic-beam scattering, 420, 467-469
  elastic, 452, 468
  energy transfer, 469
  hard-cube model, 467
  inelastic, 437, 468
  soft-cube model, 467
Atomic scattering factor, effective, 464
Atom-probe field ion microscope, 408
Attenuated total reflection, 436
Auger electron appearance potential
  spectroscopy, 393
Auger electron spectroscopy, 383, 394, 397
  electron-excited, 394
  x-ray excited, 383
Auger spectroscopy, *see* Auger electron
  spectroscopy

B

Bandwidth, 413
Boltzmann factor, 466
Bremsstrahlung, 390, 399
Brillouin zone, 420, 428, 431, 462, 464, 466
Brittle solids, deformation of in contact
  problems, 518
Brownian motion, 329
Bulk diffusion, measurement of, 307
Bulk impurity, diffusion of to surfaces, 477

Index for Volume I

A

Accumulation layer, 257
Acetylene, surface structures on Pt, 65
Activation energy
  for adsorption, 192
  for desorption, 192
Adsorption, 189-239
  dissociation in, 211
  equilibrium, 191
  experimental techniques, 222-225
  Fowler-Guggenheim Model, 219-221
  heat of, 18, 200
  Langmuir isotherm, 31, 209
  macroscopic theory, 194-202, 217-219
  monolayer model of, 27, 206-221
  multilayer, 37, 213-215
  statistical theory of, 21, 202-221
Aluminum
  clean surface, 60, 62
  substrate for Na, 64
Argon, surface structures on Nb, 5
Ashcroft potential, 91
Auger spectroscopy, estimate of surface
    coverage, 6, 223, 231
Augmented plane waves, LEED, 34
Average carrier distance from surface, 259,
    263, 264

B

Band bending, 80, 257
Band narrowing, 99
Basis (lattice), 18
Beam representation, 34, 50

Beryllium, 8
  surface structure, 60
Bethe model (phase transition), 47, 219
Bethe theory, 34
BET isotherm, 213-215
Bloch periodicity, LEED theory, 33
Bloch waves, 94
Boltzmann equation, 245
Bragg peaks, 48
Bragg-Williams method, 155
Bravais lattice, two-dimensional, 21, 38
Bulk plasmons, 115

C

Canonical partition function, 128
Carbon monoxide
  surface structures on Ni, 64
  surface structures on Mo, 65
Cesium
  surface structures on Ni, 64
  surface structures on Si, 65
Chalcogens, surface structures on Ni, 68
Chemical potential, 130
Chemisorption, 4, 50 *et seq.*
  Co on Pd, 225
  experimental results, 225-232
  S on Au, 230
  table of results, 226
Classical approximation, for space charge
    layer, 256, 265
Clausius-Clapeyron equation, 200-201, 218
Clean surfaces, 30, 56, 57
  characterization by LEED, 60
  methods of cleaning in situ, 8

Surface torque, 148
Surface transport
   via space charge layer, 266
   via surface states, 249
   types of, 242

U

Unit mesh
   definition, 19
   substrate, 28
   surface, 28

V

Vanadium, surface structures, 60
Vicinal surfaces, 141, 212-213

T

Tellurium
   surface structures on Ag, 65
   surface structures on Ni, 64, 69
Thermal roughening, 153
Thermal vibrations, LEED, 51-56
Thickness of the interfacial region, 152
Titanium, surface structure, 60
T operator, LEED theory, 17, 18, 40
   symmetry properties of, 25-28
Transfer matrix (**Q** matrix), 41, 96
Transition probabilities, 168
Transition region, 192, 204
Translation group, two-dimensional, 19
Tungsten
   adsorption sites, 6
   method of cleaning in situ, 8
   substrate for O, 65
   surface structure, 60
Two-dimensional Ising model, 173

W

Wigner interpolation formula, 85
Work function, 87
   Na, 90
   Si, 88
Wulff's theorem, 126

X

Xenon, surface structures on Nb, 5

Z

Zero point vibration, 140
Zinc, surface structures, 60
Zinc oxide, clean surface, 250